TEACHER'S RESOURCE

SERIES EDITOR: BRIAN SEAGER

OCR GCSE
MATHEMATICS

STAGES

9
10

GRADUATED ASSESSMENT

SECOND EDITION

- Howard Baxter
- Michael Handbury
- John Jeskins
- Jean Matthews
- Mark Patmore

HODDER
EDUCATION
PART OF HACHETTE LIVRE UK

Hodder Headline's policy is to use papers that are natural, renewable and recyclable products and made from wood grown in sustainable forests. The logging and manufacturing processes are expected to conform to the environmental regulations of the country of origin.

Orders: please contact Bookpoint Ltd, 130 Milton Park, Abingdon, Oxon OX14 4SB. Telephone: (44) 01235 827720. Fax: (44) 01235 400454. Lines are open 9 a.m. to 5 p.m., Monday to Saturday, with a 24-hour message-answering service. Visit our website at www.hoddereducation.co.uk.

© Howard Baxter, Michael Handbury, John Jeskins, Jean Matthews, Mark Patmore, Brian Seager, Eddie Wilde, 2007
First published in 2007 by
Hodder Education,
part of Hachette Livre UK,
338 Euston Road
London NW1 3BH

Impression number   10 9 8 7 6 5 4 3
Year   2012 2011 2010 2009 2008

Cover photo © Andy Sacks/Photographer's Choice/Getty Images
Typeset in 10pt TimesTen by Pantek Arts, Maidstone, Kent.
Printed in Great Britain by Hobbs The Printers, Totton, Hants.

A catalogue record for this title is available from the British Library.

ISBN: 978 0340 91600 1

Stage 9 Contents

STAGE

9

Introduction

About the Student's Book

This course has been written especially for students following OCR's Modular Specification C, Graduated Assessment (J516 and J517) for GCSE Mathematics. The Student's Book covers the complete specification for Stages 9 and 10.

- Each chapter is presented in a way which will help students to understand the mathematics, with straightforward explanations and worked examples covering every type of problem.
- At the start of each chapter are two lists: one of what your students should already know before they begin and the other of topics they will be learning about in that chapter.
- 'Activities' offer a more interesting approach to the core content, giving opportunities for students to develop their skills.
- 'Challenges' are rather more searching and are designed to make students think mathematically.
- There are plenty of exercises for students to work through to practise skills.
- Some questions are designed to be done without a calculator, so that students can practise for the non-calculator sections of the examination papers.
- Look out for the 'Exam tips' – these give advice on how students can improve their performance in the module tests, direct from the experienced examiners who have written this book.
- At the end of each chapter there is a short summary of what has been learned.
- Finally, there are 'Revision exercises' at intervals throughout the book to help students revise all the topics covered in the preceding chapters.

About this book

The Teacher's Resource is designed to save you valuable planning time. A CD-ROM is included, containing everything in this book. It may be customised for use in your school.

- At the start of each chapter in the book is the statement from the specification which is covered by the chapter in the Student's Book, together with a list of the objectives for the chapter and a list of the prior knowledge required.
- The 'General notes' give specific advice on teaching the topics covered in the chapter.
- Answers are given to all the Activities and Challenges, as well as ideas for how best to use these in class, providing plenty of ideas for active and engaging lessons.
- Photocopiable 'Speed-up sheets' are provided to ensure students can concentrate on practising their mathematical skills rather than spend time copying from their student's books.
- A formulae sheet and mathematical grids are also provided and can be photocopied.
- Answers to all the exercises in the Student's Book are also provided.

Other components in the series

- Personal Tutor CD-ROM
  The course has bespoke digital resources. The Personal Tutor CD-ROM uses the most up-to-date technology to provide audio-visual presentations that can be used as a whole-class teaching tool or for individual personal tuition. The CD-ROM contains step-by-step solutions to worked examples covering all the key concepts from the Student's Book. This allows students to revise difficult concepts and consolidate knowledge and provides a virtual teaching environment for independent study. The CD-ROMs are networkable within the purchasing institution.

STAGES
9/10

■ A Homework Book

Although there are plenty of exercises throughout the Student's Book for your students to work through, it is not always possible for them to work from their textbooks at home. The separate Homework Book contains a parallel exercise for every exercise in the Student's Book, providing you with ready-made homeworks. Included with the Homework Book is a student edition of the Personal Tutor CD-ROM. This will help students if they miss a lesson or need a reminder of something taught in class. You will find the answers to the Homework Book in the Higher Assessment Pack.

■ An Assessment Pack

There are two Assessment Packs: one for Foundation Tier (Stages 1 to 7) and one for Higher Tier (Stages 6 to 10). Each contains detailed information about the examination requirements. In addition there are revision exercises, practice module papers and a practice terminal paper to help your students prepare for the examination, together with answers to all of these. Some of the questions in the examination will offer little help to get started. These are called 'unstructured' or 'multi-step' questions. Instead of the question having several parts, each of which helps candidates to answer the next, they have to work out the necessary steps to find the answer. There are examples of this kind of question in the Assessment Pack.

■ An Interactive Investigations CD-ROM

This contains whole-class presentations and individual activities. It will help you to integrate ICT into your lessons.

STAGES
9/10

Graduated Assessment for OCR GCSE Mathematics © Hodder Murray 2007

Checking answers

N9.2

Check the order of magnitude of a compound calculation using estimation methods, including rounding numbers of any size to one significant figure and simplifying calculations using standard index form, without the use of a calculator.

H2/3h, 3m, 4b

Objectives

- Check answers to calculations by rounding to 1 significant figure
- Simplify calculations using standard form

Prior knowledge

- Interpret significant figures
- Write numbers in standard form

Equipment needed

None

General notes

In this chapter students encounter compound calculations involving ordinary and standard form numbers. The method of rounding to 1 significant figure is required throughout the exercises.

In the first instance, allow students to check their answers by evaluating the original expression with a calculator. This gives them the opportunity to practise important calculator methods, including the use of standard form, and enables them to check whether their answer is of the correct order. Once students become more confident in their approach, the use of calculators should be eliminated.

Notes on the tasks

Activity 1 (page 2)

All the calculations are incorrect. Possible reasons include the following.

a) odd $\times$ odd = odd
b) Multiplying by a number between 0 and 1 makes the original number smaller.
c) $4000 \div 100 = 40$, so the magnitude of the answer is incorrect.
d) $9 \times 7 = 63$, so the last digit should be 3.
e) Dividing by a number between 0 and 1 makes the original number larger.
f) $6000 \times 1000 = 6\,000\,000$, so the magnitude of the answer is incorrect.

Challenge 1 (page 4)

Check students' calculations and estimates.

STAGE

9

Algebraic manipulaton

A9.3

Manipulate algebraic expressions by expanding the product of two linear expressions, by taking out common factors and by cancelling common factors in rational expressions; factorise quadratic expressions, including the difference of two squares; solve quadratic equations of the form $ax^2 + bx + c = 0$ by factorisation.

H2/5b, 5k

Objectives

- Expand brackets such as $(2x + 3y)(3x + 2y)$
- Simplify algebraic fractions such as $\frac{3x^2y^4}{6xy^2}$ and $\frac{x^2 + x}{x^2 + 3x + 2}$
- Factorise expressions such as $20a^2b^3 + 5ab$ and $8x^2 + 4x - 12$
- Recognise and factorise expressions such as $x^2 - y^2$ and $4a^2 - 9b^2$
- Solve quadratic equations of the form $ax^2 + bx + c = 0$ by factorisation

Prior knowledge

- Expand brackets and manipulate simple algebraic expressions

Equipment needed

None

General notes

Students can expect to find a number of questions on manipulative algebra in the examination papers. The work in this chapter must therefore be covered in detail. It will also be a good foundation for students going on to study mathematics at A Level.

Students commonly give the answer to expressions of the type $(a + b)^2$ as $a^2 + b^2$. They should be reminded that this type of expression will always produce three terms. Only the difference of two squares produces just two terms.

Simplification of algebraic expressions often involves indices. It is worth encouraging students to learn the four laws of indices given in the Student's Book.

Factorisation includes taking out common factors from expressions and factorising into two brackets. Encourage students to check their answers by multiplying out the two brackets to get back to the original expression.

You may want to recap the factorisation of quadratic expressions in which the coefficient of x^2 is 1 before going on to expressions in which it is not 1. Emphasise that the first step with the latter type of expression is to check whether there is a common factor for the whole expression: if there is, taking it out will simplify the quadratic, making the factorisation more straightforward.

Two basic rules for factorising quadratic expressions are worth mentioning.

- When there is no number term, look to take out x as a common factor.
- When there is no x term, look to use the difference of two squares.

Cancelling algebraic fractions includes cancelling numbers, letters or brackets. Remind students that it is incorrect to cancel part of an expression. They must cancel only factors of the entire numerator and denominator.

Solving quadratic equations of the form $ax^2 + bx + c = 0$ extends the solution of quadratic equations in the same way as factorising quadratic expressions was extended earlier in the chapter. When a common factor has been taken out before factorisation, emphasise the fact that dividing zero by anything results in zero, so the factor does not form part of either of the solutions to the equation.

Notes on the tasks

Challenge 1 (page 7)

a) (i) $(4x + 3)$ m   (ii) $(x + 2)$ m
b) $(4x + 3)(x + 2) = (4x^2 + 11x + 6)$ m^2

Challenge 2 (page 8)

a) (i) $(5x - 5)$ m   (ii) $(3x - 2)$ m
b) $(5x - 5)(3x - 2) = (15x^2 - 25x + 10)$ m^2

Challenge 3 (page 10)

Check students' work.

Challenge 4 (page 17)

a) $(4x + 3)(2x + 1)$
b) $(5x + 4)(3x - 2)$
c) $(4x + 5)(2x - 3)$
d) $(3x - 7)(2x - 5)$

Challenge 5 (page 23)

$$325 = \tfrac{1}{2}n(n + 1)$$
$$650 = n^2 + n$$
$$n^2 + n - 650 = 0$$
$$(n - 25)(n + 26) = 0$$
$$n = 25 \text{ or } ^-26$$

Since S is the sum of the first n positive integers, $n = 25$.

This challenge involves dealing with a fraction and interpreting the context of the question to choose the correct answer.

STAGE
9

Graduated Assessment for OCR GCSE Mathematics © Hodder Murray 2007   3

3 Proportion and variation

A9.2

Form and use equations to solve word and other problems involving direct and inverse proportion (for example, $y \propto x$, $y \propto x^2$, $y \propto 1/x$, $y \propto 1/x^2$) including relating algebraic solutions to graphical representations of the equations.

H2/3I, 5h

Objectives

- Solve problems involving direct and inverse proportion

Prior knowledge

- Find and use multipliers
- Manipulate simple formulae
- Substitute numbers into simple formulae

Equipment needed

None

General notes

The chapter begins with an intuitive approach, assuming, without precisely defining, direct and inverse linear proportion in terms of going up and going down. This is then developed to cover other forms of proportion.

The unitary method has been avoided in favour of an approach based on scale factors. Links with other areas of mathematics, such as similar shapes, areas and volumes, can usefully be developed.

The use of the $\propto$ symbol is a convenient shorthand and appears in the specification and on examination papers. All the forms of proportion that students need to know are covered in this chapter. More difficult questions may involve a combined variation with three variables: these can be answered by tackling the question in two stages. Challenge 1 is an example of this type of question.

Notes on the tasks

Challenge 1 (page 32)

Mass (m) is proportional to length (l) so $m \propto l$.

Mass (m) is also proportional to the square of the diameter (d) so $m \propto d^2$.

Mass of 1 km of wire with a diameter of 2 mm
$$= 31 \cdot 5 \times 2 \times \tfrac{4}{9}$$
$$= 28 \, \text{kg}$$

Indices 4

N9.3

Use fractional, negative and zero powers in simplifying numerical expressions, including using inverse operations.

H2/3a, 3g, 5d

Objectives

- Use the laws of indices
- Deal with negative and fractional indices

Prior knowledge

- Work out numbers using positive and negative indices
- Use the basic laws of indices
- Round to a given number of significant figures
- Know the meaning of the words *power*, *prime* and *factor*

Equipment needed

2 mm graph paper

Speed-up sheets available

4.1

General notes

This chapter introduces the idea of fractional indices. Students should already have used positive, zero and negative indices, but they are revised here.

In the text, the fact that $a^{\frac{1}{3}}$ is the same as $\sqrt[3]{a}$ is shown, but no proof is given of the general law.

Students also need to be able to work out powers and roots using a calculator, and should familiarise themselves with the use of the appropriate keys.

The second part of the chapter looks at the laws of indices. These have been introduced earlier in the course, and are now extended to include fractional indices. It is important that students deal with both numbers and letters correctly. A common error in an expression such as $(2a^2)^4$ is to multiply the 2 by the 4 and get $8a^8$: ensure students realise that the correct answer is $(2a^2)^4 = 2^4 \times a^{2\times4} = 16a^8$.

The writing of numbers as a product of their prime factors and using indices was covered in Stage 7 but is also included here as an application of the laws of indices.

Notes on the tasks

Activity 1 (page 38)

$2^3 = 8$, $\left(\frac{1}{2}\right)^{-2} = 4$, $1000^{\frac{2}{3}} = 100$, $64^{\frac{1}{2}} = 8$, $10^2 = 100$, $64^{\frac{1}{3}} = 4$, $(0.05)^{-1} = 20$, $8^{\frac{2}{3}} = 4$, $4^{\frac{1}{2}} = 2$, $64^{\frac{2}{3}} = 16$, $64^{\frac{1}{6}} = 2$, $7^0 = 1$, $36^{\frac{1}{2}} = 6$, $64^{\frac{5}{6}} = 32$

Notice that not all the whole numbers are used, and some are used more than once.

STAGE

9

Challenge 1 (page 39)

a)

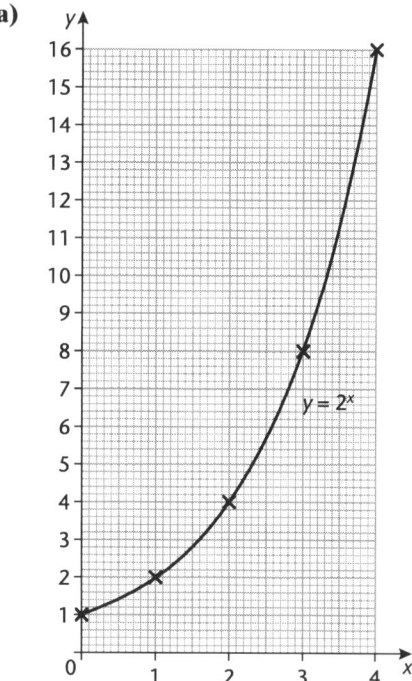

b) $2^{\frac{1}{2}} \approx 1{\cdot}4 \qquad 2^{\frac{1}{3}} \approx 1{\cdot}25 \qquad 2^{\frac{3}{2}} \approx 2{\cdot}8$

$$\left(2^{\frac{1}{2}}\right)^2 = 2^{\frac{1}{2}\times 2} = 2^1 = 2$$

This suggests that $2^{\frac{1}{2}} = \sqrt{2}$.

$\sqrt{2} = 1{\cdot}41$ (to 2 d.p.)

c) $\left(2^{\frac{1}{3}}\right)^3 = 2^{\frac{1}{3}\times 3} = 2^1 = 2$

This suggests that $2^{\frac{1}{3}} = \sqrt[3]{2}$.

$\sqrt[3]{2} = 1{\cdot}26$ (to 2 d.p.)

d) $\left(2^{\frac{3}{2}}\right)^2 = 2^{\frac{3}{2}\times 2} = 2^3$

This suggests that $2^{\frac{3}{2}} = \sqrt{2^3}$.

$\sqrt{2^3} = 2{\cdot}83$ (to 2 d.p.)

STAGE
9

Rearranging formulae

A9.1

Rearrange harder formulae, including cases where the subject appears twice, or where a power of the subject appears.

H2/5g

Objectives

- Rearrange harder formulae with fractional terms and indices
- Rearrange formulae where the subject occurs more than once

Prior knowledge

- Factorise simple expressions
- Rearrange simple formulae
- Expand brackets and manipulate simple algebraic expressions

Equipment needed

None

General notes

This extends the rearrangement of formulae to include more complicated examples where the subject occurs more than once or where a power or root of the subject occurs, although rearranging formulae where a root of the subject occurs is at the limit of the course.

It is useful to stress that many candidates make errors in examinations when rearranging formulae, but if they show their working they may earn some marks for any steps carried out correctly.

STAGE
9

6 Arcs, sectors and volumes

S9.3

Solve problems involving the lengths of arcs, areas of sectors and the volume of pyramids, cones and spheres.

H3/2i, 4d

Objectives

- Find the length of an arc and the area of a sector of a circle, using the appropriate formula
- Find the volume of cones, spheres and pyramids, using the appropriate formula

Prior knowledge

- Find the circumference and area of a circle
- Find the volume of a prism
- Rearrange formulae
- Use Pythagoras' theorem and trigonometry

Equipment needed

Models of three-dimensional shapes (optional)

General notes

In this chapter there are clear opportunities to revise and develop topics covered earlier. There are also opportunities to extend the work beyond the strict confines of the specification. One such opportunity is to introduce radians, which students may be familiar with from their calculator modes,

and will need in A-Level study of physics or mathematics. Exercise 6.1 questions **1(i)** and **4(f)** could act as a starting point for discussion. It is worth making it clear, of course, that this knowledge is not required for GCSE.

For those who find three-dimensional work difficult, models of pyramids, cones and other solids can help enormously in interpreting the context.

Students may well need assistance with some of the more complex problems. However, presenting a difficult question as an enjoyably tough challenge rather than as something to be worried over may help. Practice in coping with unstructured questions, which this chapter provides, is also useful.

Notes on the tasks

Activity 1 (page 46)

a) Three
b) Area = $26 \cdot 2 \, \text{cm}^2$ (to 1 d.p.),
   arc length = $10 \cdot 5 \, \text{cm}$ (to 1 d.p)
c) Perimeter = arc length + 2 × radius

This task encourages discussion about sectors as fractions of circles. It is important that students fully grasp that the sector shown is one third of the full circle and see that the area of the sector can therefore be expressed as $\frac{1}{3}\pi r^2$, rather than simply dividing the area of the full circle by 3. Extending the activity to consider other sectors, including angles which are not factors of 360°, will encourage this.

Challenge 1 (page 52)

a) $2 \cdot 61 \, \text{cm}$
b) $5 \cdot 66 \, \text{cm}$ (to 2 d.p.)
c) $9 \cdot 92 \, \text{cm}$ (to 2 d.p.)

Challenge 2 (page 56)

$131 \, \text{cm}$ (to 3 s.f.)

Upper and lower bounds

7

N9.1

Use calculators or written methods to calculate the upper and lower bounds of calculations, particularly in the context of measurement.

 H2/3q

Objectives

- Calculate the upper and lower bounds of measurements and recognise the effect of these on calculations

Prior knowledge

- Interpret a measurement given to a stated degree of accuracy

Equipment needed

Measuring equipment (optional)

General notes

A short introductory exercise on practical measurement could be used to set the scene for this topic. Discussion of the accuracy achievable with different measuring aids and the consequent effect on calculations using those measurements gives relevance, especially if real-life measurements from recent experimental work in other subjects such as geography, science or food technology can be used.

The examples given in the text and concepts used in the exercises include speed and density as well as basic areas and volumes.

STAGE

9

8 Similarity and enlargement

S9.4

Understand and use the effect of enlargement on length, area and volume of shapes and solids, including the use of negative scale factors.

H3/3c, 3d

Objectives
- Find the volume and surface area of similar figures
- Enlarge shapes using a negative scale factor

Prior knowledge
- Find volumes and areas
- Use scale factors

Equipment needed
Squared paper, multilink cubes (optional)

Speed-up sheets available
8.1

General notes

Practical work will help to reinforce the concepts about comparing the volumes or areas of similar shapes. An investigative approach could successfully be used to introduce the topic.

The work on enlargement with a negative scale factor completes the development of this concept.

Notes on the tasks

Challenge 1 (page 70)
a) Students should predict that, generally speaking, similar birds are larger in colder climates. The scale factor for volume (greater volume, greater food intake) is greater than that for surface area, so larger birds have more energy to combat heat loss.
b) A similar argument. If a small aircraft were to be enlarged and retain the ability to fly, either the wing area must be increased by more than the scale factor or the volume (or mass) reduced below the scale factor. Of course, if model aircraft are considered, the actual speeds achieved are also scaled down.

Challenge 2 (page 72)
a) Rotation of 180° about the point chosen as the centre of enlargement.
b) Depending on the triangle and the centre of enlargement, reflection is possible, although only the complete shape is mapped, not corresponding points. For example, here an isosceles triangle has been enlarged by a scale factor of ⁻1, with centre of enlargement O: a reflection in the line drawn through O would give the same effect, but the points A_1 and C_1 would be transposed.

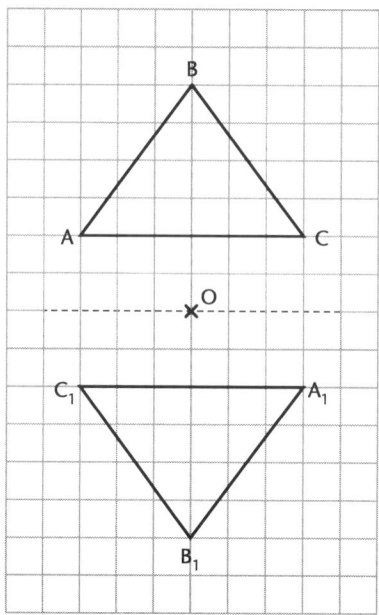

Graduated Assessment for OCR GCSE Mathematics © Hodder Murray 2007

D9.1

Solve structured problems involving the addition or multiplication of two probabilities.

H4/4g

Objectives

- Use the addition rule for mutually exclusive events,
  P(A or B) = P(A) + P(B)
- Use the multiplication rule for independent events,
  P(A and B) = P(A) × P(B)
- Find the probabilities of dependent events
- Recognise conditional probability

Prior knowledge

- Draw and use a tree diagram
- Understand and use the terms *independent* and *mutually exclusive*

Equipment needed

None

Speed-up sheets available

9.1

General notes

The chapter starts with revision of the addition and multiplication rules, which were covered in Stage 8. It is important that students grasp when to use which rule. The examples show how to answer problems without the use of a tree diagram. These are, however, powerful tools for clarifying problems and you should emphasise to students that they may well find it helpful to draw a tree diagram.

The chapter then moves on to conditional probability. There is no need for the formal approach to this topic that would be required at A Level. Students should learn to recognise when events are independent and when they are dependent. The key point to grasp when working with conditional probability is that, although the probabilities change for the second event (and subsequent events), the basic rules remain the same. Once students have understood this, they should have no more difficulty with dependent events than with independent events.

You may wish, with more able students who expect to go on to study mathematics at A Level, to introduce the formal notation for conditional probability.

The probability of event B happening, given that event A has already happened is written as P(B | A).

The multiplication rule then becomes
P(A and B) = P(A) × P(B | A).

The tree diagram for a simple two-event problem looks like this.

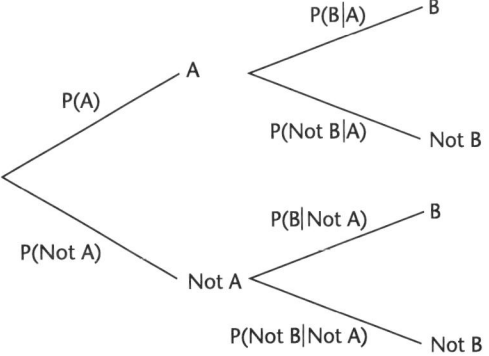

In the Student's Book, tree diagrams are used for conditional probability. At this level, however, the instruction to draw a tree diagram is often not given, the choice of method being left to the student. Less able students are strongly advised to draw a tree diagram in order to ensure they cover all possible combinations of events, whether for independent or dependent events.

STAGE

9

Working in two and three dimensions

S9.2

Use Pythagoras' theorem and trigonometrical relationships in 3-D contexts, including using 3-D coordinates and finding the angles between a line and a plane; use Pythagoras' theorem to find the length AB given the points A and B in 2-D.

H3/2f, 2g, 3e

Objectives

- Calculate the distance between two points
- Use right-angled triangles to find lengths and angles in three dimensions
- Find angles between lines and planes

Prior knowledge

- Use coordinates in two and three dimensions
- Apply Pythagoras' theorem
- Use the convention for labelling sides and angles in a triangle
- Use trigonometry in right-angled triangles

Equipment needed

Squared paper (optional), models of three-dimensional shapes (optional)

General notes

Before making a start on this chapter it is a good idea to revise what students have already learned about coordinates in two and three dimensions, Pythagoras' theorem and trigonometry in right-angled triangles.

The first section of the chapter is about finding the distance between two points in two dimensions. This is a useful introduction to identifying the relevant right-angled triangle. Encourage students to draw a diagram showing the triangle and mark on the lengths. Emphasise the need to work out the lengths from the coordinates and not by counting squares: this will avoid errors when the scale is not one division to one unit.

The chapter then moves on to problems in three dimensions. Time spent discussing solid figures and their main features is useful. You may find the use of solid or skeleton models helpful here.

Stress that a good way to start tackling problems of this sort is with a simple diagram of the three-dimensional shape. You may find it helpful to give these instructions on how to draw it.

- Draw vertical lines parallel to the edges of the paper.
- Draw hidden edges using dashed lines.
- Make parallel edges look parallel.
- Don't worry about making the diagram look in perspective.

Once students have drawn a diagram, the second step is to identify the right-angled triangle containing the required side or angle. You may need to point out that a line perpendicular to a plane is perpendicular to every line in that plane. In the diagram opposite, for example, RC is perpendicular to DC, BC and AC since RC is perpendicular to the plane ABCD.

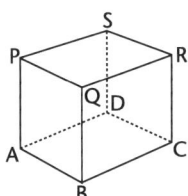

A solid understanding of this fact is particularly useful because a right angle in a three-dimensional sketch often does not look like a right angle.

The third step is to draw another diagram, this time showing only the relevant right-angled triangle. Students should label any known lengths and angles, and mark any right angles.

Questions involving three-dimensional shapes are often multi-step, requiring students to find an intermediate length or angle before the required length or angle can be calculated. Some of the questions in Exercise 10.2 are of this type.

The topic is then developed in two ways. The penultimate section of the chapter is about finding the length between two points in three dimensions. Students met three-dimensional coordinates in Stage 7 but they may need reminding of the convention that the x–y-plane is horizontal and the z-axis 'vertical'. Students should be able to relate finding the distance between two points in three dimensions to finding lengths in a cuboid quite readily.

The final section of the chapter covers finding the angle between a line and a plane. The angle is sometimes difficult to spot, particularly if the plane is not at the bottom of the shape. Encourage students to turn the diagram around so that the required plane is at the bottom. Some initial practice at this will avoid problems later.

More formally, the angle between a line and a plane is defined as the angle between the line and the projection of the line on to the plane.

STAGE
9

Histograms

D9.2

Draw and interpret histograms for grouped data: understand frequency density.

H4/4a, 5d

Objectives

- Construct and interpret histograms

Prior knowledge

- Calculate the mean of a set of data
- Calculate the mean for grouped data
- Understand and use the symbols $\leqslant$ and $<$

Equipment needed

2 mm graph paper

Speed-up sheets available

11.1, 11.2

General notes

This chapter involves the drawing and interpreting of histograms. Towards the beginning of the chapter, the differences between histograms and bar charts are summarised in four bullet points: it is important that students appreciate these differences, and the histogram that follows should be used to illustrate them. This will establish when it is appropriate to draw a histogram and how they can be used to find information. Once students have grasped

the need for a different approach when dealing with groups of unequal width, the main difficulty is in mastering the calculations involved.

Finding an estimate of the mean of grouped data was covered in Stage 7 and is revised here.

Notes on the tasks

Activity 1 (page 102)

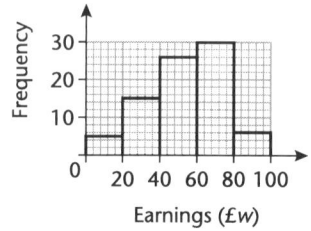

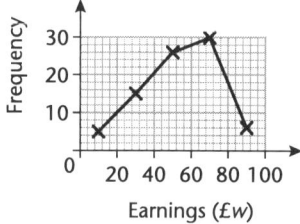

Students' ideas about how to deal with the second set of data, where the groups are of unequal width, are likely to include a bar chart like this one.

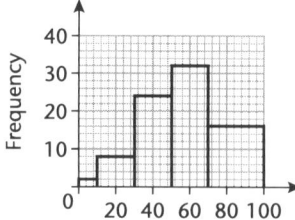

Although the heights of the bars show the frequency, students should realise from looking at this chart that our eyes use area to interpret bar graphs. Thus the wider groups look larger than they should, given the data they represent.

Graduated Assessment for OCR GCSE Mathematics © Hodder Murray 2007

Circle properties

S9.1

Use and prove angle and tangent properties of circles, including the alternate segment theorem.

H3/2h

Objectives

- Understand and use the seven geometrical properties of a circle
  - The angle subtended by an arc at the centre of a circle is twice the angle subtended at the circumference
  - The angle subtended at the circumference in a semicircle is a right angle
  - Angles in the same segment of a circle are equal
  - The opposite angles of a cyclic quadrilateral add up to 180°
  - The perpendicular from the centre to a chord bisects the chord
  - The two tangents to a circle from an external point are equal
  - The angle between a tangent and a chord is equal to the angle in the alternate segment

Prior knowledge

- Know that the sum of the angles in a triangle equals 180°
- Know that the exterior angle of a triangle equals the sum of the interior opposite angles
- Know that the sum of the angles on a straight line equals 180°
- Know the meaning of the term *congruent*
- Know the meaning of circle terms such as *arc*, *sector* and *segment*
- Know that the tangent at any point on a circle is perpendicular to the radius at the point

Equipment needed

Pairs of compasses (optional)

General notes

This chapter is intended to form a basis for further work on proof and reasoning. Euclidean geometry is used to provide the opportunities for the development of coherent and logical arguments and deductions. The angle facts that students need to know are deduced and students are expected to work through these deductions. Alternatively, you could get students to produce their own proofs first for some of the theorems.

Emphasise that, when answering questions in the exercises, students are expected to express their own deduction/reasons for the answers they give. Note that there may be different routes to the answers in some cases so you will need to check the reasons given by each student. Answers which are not supported by working and reasons should not be awarded full marks.

STAGE

9

13 Straight-line graphs

A9.4

Find gradients of straight lines perpendicular to each other and write equations of straight lines in the form $y = mx + c$.

H2/6c

Objectives

- Write the equation of a straight line in the form $y = mx + c$
- Find the gradient of straight lines which are perpendicular to each other

Prior knowledge

- Draw a straight-line graph when given coordinates or the equation of the line
- Know that $y = mx + c$ represents a straight line
- Know that m represents the gradient of the line and c represents the y-intercept
- Find the gradient of a straight-line graph
- Recognise when lines are parallel

Equipment needed

Graph-drawing software or 2 mm graph paper

Speed-up sheets available

13.1, 13.2, 13.3

General notes

In Stage 8 students learned how to find the gradient of a straight line and about the y-intercept of a line. It is a small step from understanding these concepts to being able to write the equation of a straight line in the form $y = mx + c$. Students need to be able both to find the equation of a line when given the graph and to know what the line represented by a given equation will look like on a graph.

Notes on the tasks

Activity 1 (page 132)

a) (i)

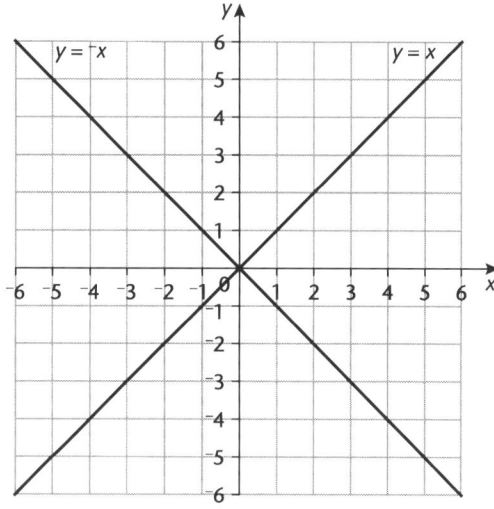

(ii)

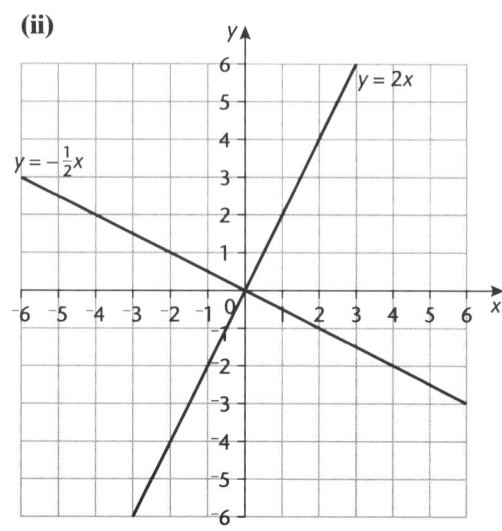

Graduated Assessment for OCR GCSE Mathematics © Hodder Murray 2007

(iii)

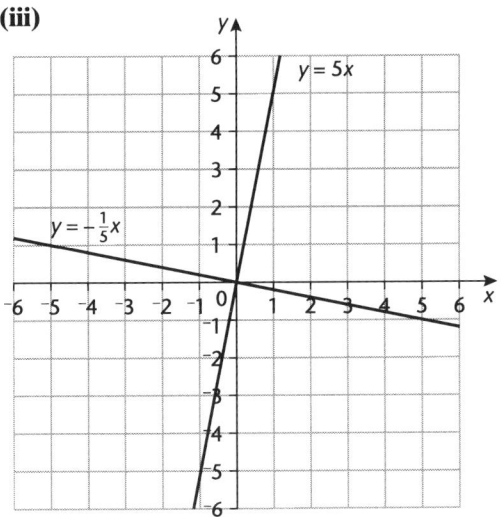

(iv)

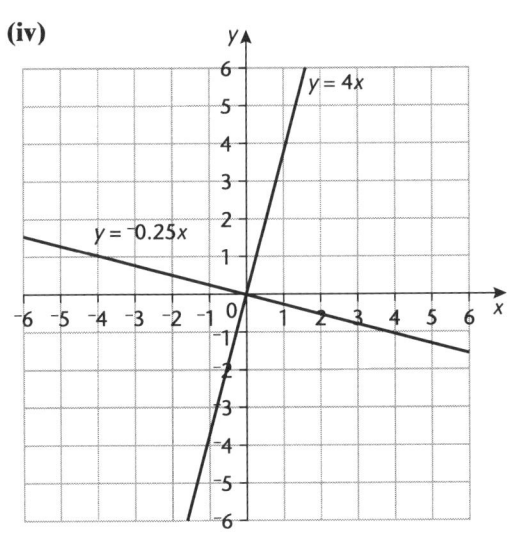

(v)

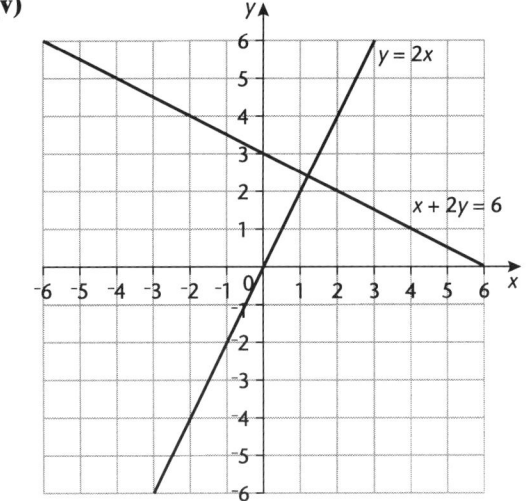

b) The lines are perpendicular.
The gradient of one line is the negative reciprocal of the other line *or* the product of the gradients is ⁻1 *or* an equivalent statement.

Note that students need to use the same scale on both axes in order to see that the lines are perpendicular.

STAGE
9

14 Surveys and sampling

D9.3

Select a representative sample from a population using random and stratified sampling; criticise sampling methods.

H4/2d

Objectives

- Conduct more sophisticated surveys
- Understand sampling
- Understand random and stratified sampling and when to use them
- Recognise bias

Prior knowledge

- Analyse data, for example, calculate the mean, median, mode and range
- Design questionnaires
- Carry out pilot surveys
- Display information and results

Equipment needed

Coloured counters (optional)

General notes

This chapter builds on knowledge and techniques gained from earlier work. Remind students as they work through this chapter that they need to be able to present and analyse data, explaining and giving reasons for what they have done; if appropriate, to use summary statistics to make relevant comparisons and draw conclusions; to relate any findings to starting hypotheses; and to comment on the validity of any inferences drawn, considering such aspects as sample size and bias.

Notes on the tasks

Activity 1 (page 136)

Check students' answers.

This helps students analyse work they have done previously. The investigation does not necessarily have to be one they did in their mathematics lessons as this work is equally applicable to other subjects.

Activity 2 (page 138)

Check students' answers.

For further work on sample size, as a department you could take some information (height, for example, or hours spend on homework) about each student in the year group, put this into a spreadsheet and calculate the average. Then repeat Activity 2 using different sized samples.

Activity 3 (page 143)

Check students' answers.

This helps to reinforce the method of simple random sampling, which is vital. It also illustrates further the effect of sample size.

Activity 4 (page 146)

Check students' answers.

This provides some ideas for putting the ideas in this chapter into practice.

Graduated Assessment for OCR GCSE Mathematics © Hodder Murray 2007

Stage 9 Speed-up sheets

Challenge 1 (page 39)

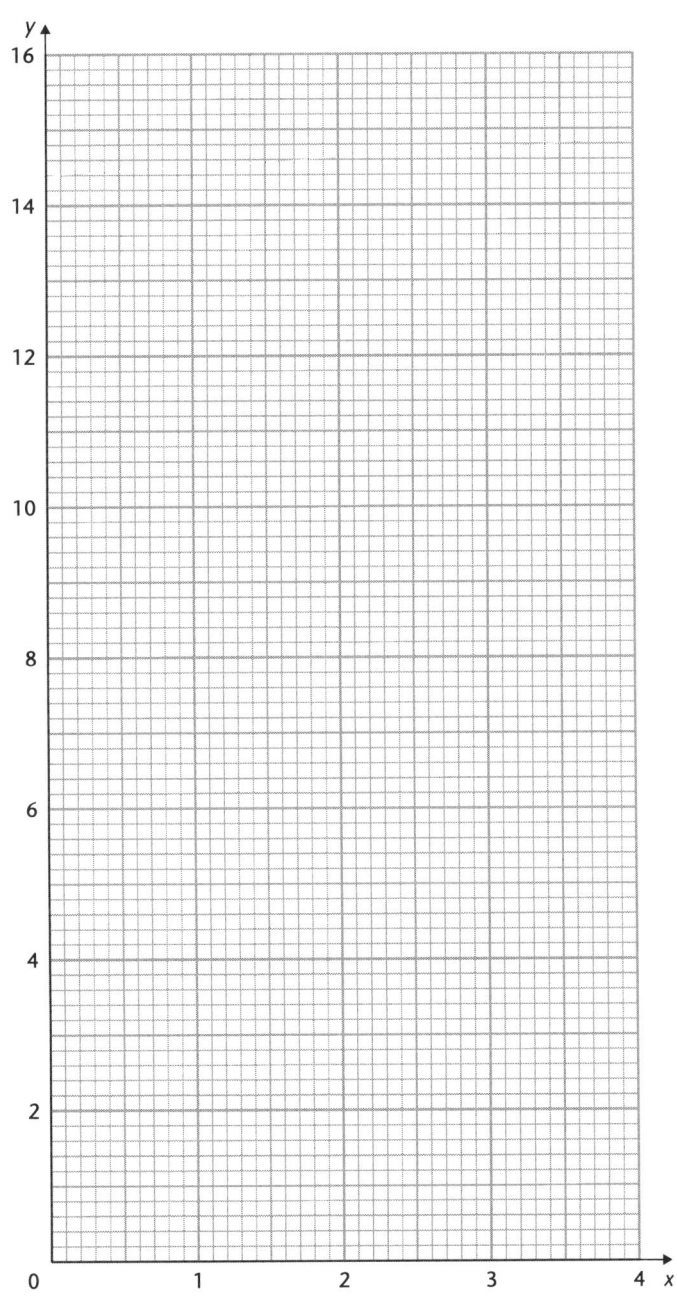

STAGE
9

Exercise 8.2 (page 71)

1

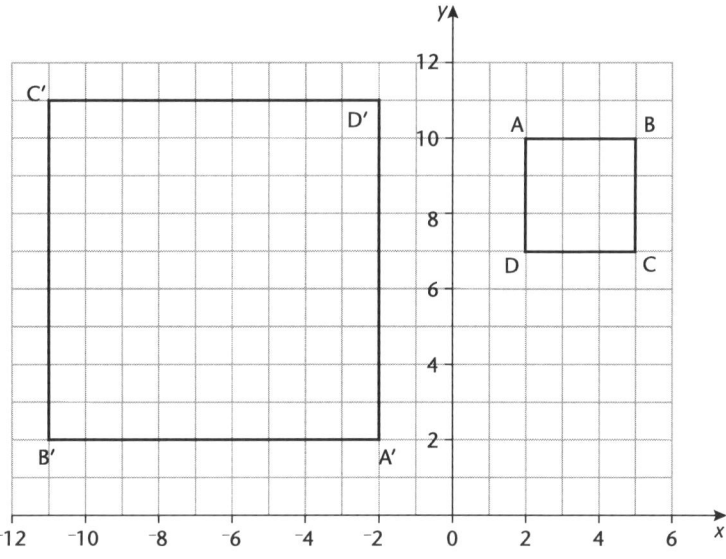

2

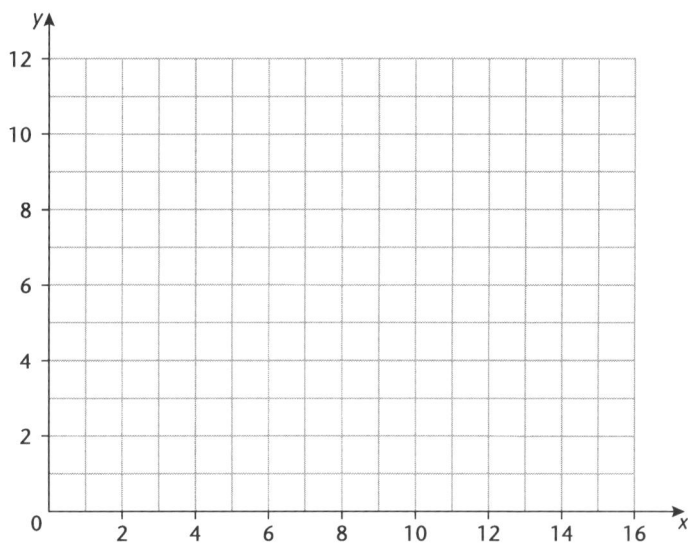

3

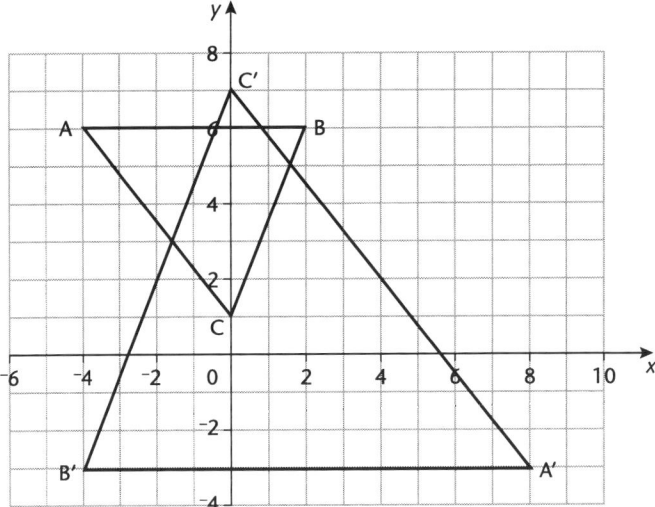

4

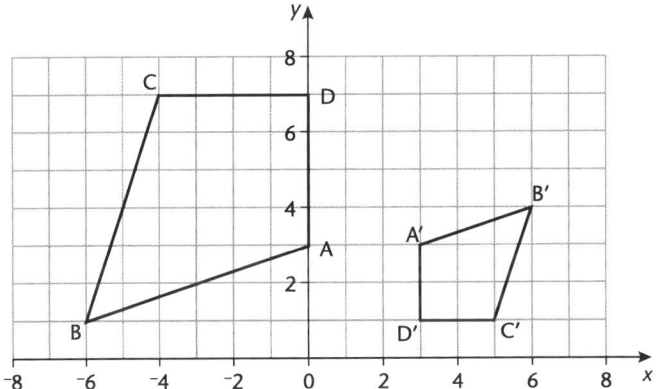

SPEED-UP SHEET 9.1

Exercise 9.2 (page 81)

5

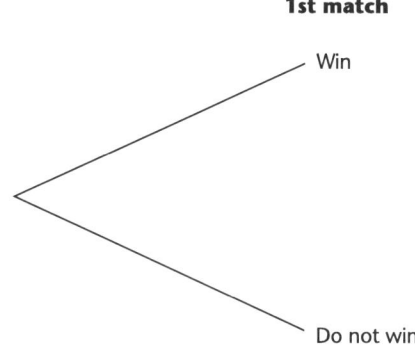

1st match

Win

Do not win

Graduated Assessment for OCR GCSE Mathematics © Hodder Murray 2007

Revision exercise C1 (page 100)

4

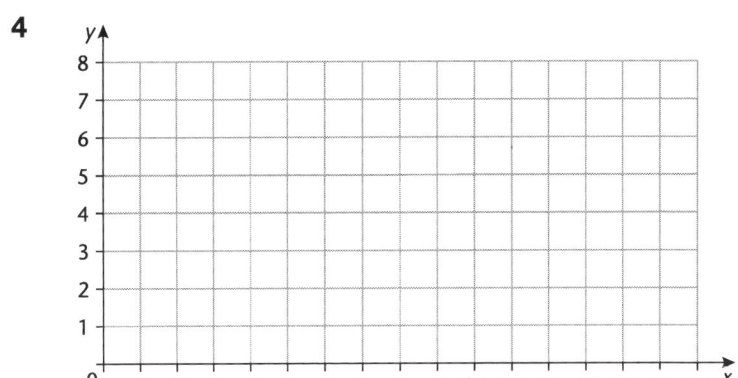

5

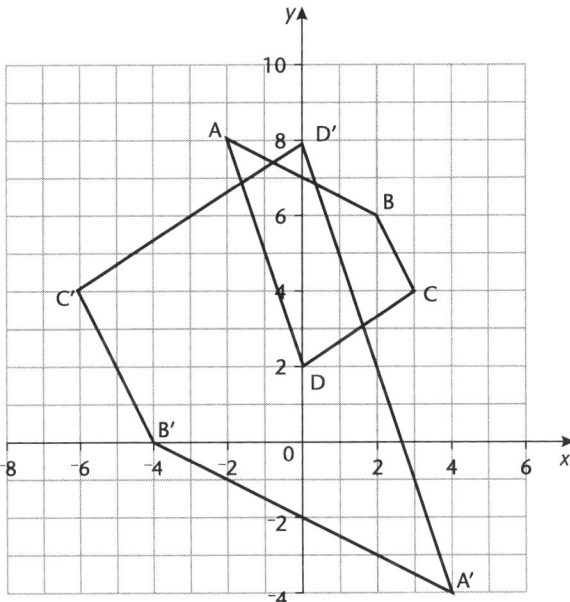

STAGE
9

Speed-up sheets

23

Activity 1 (page 102)

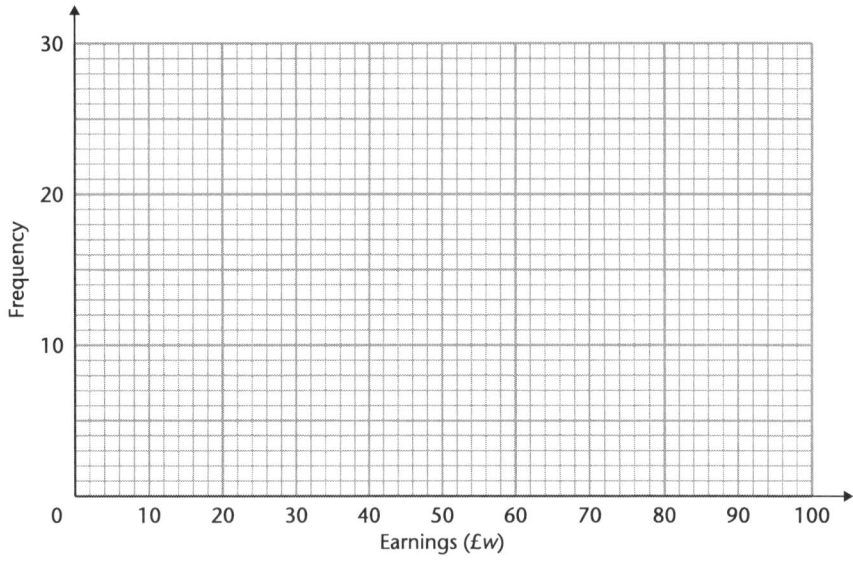

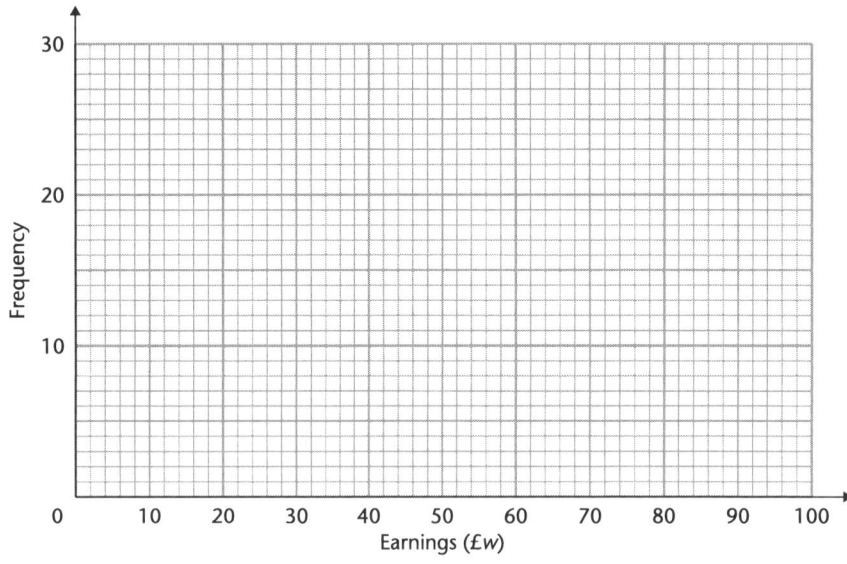

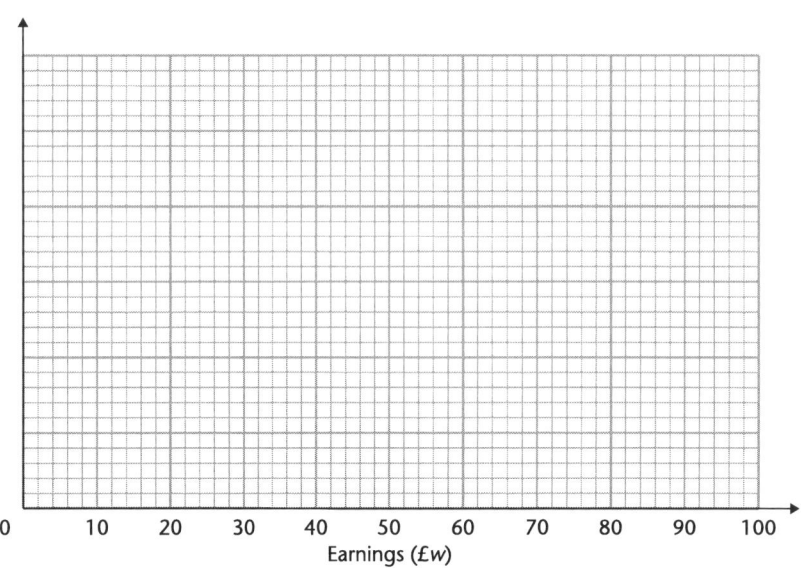

Graduated Assessment for OCR GCSE Mathematics © Hodder Murray 2007

Exercise 11.1 (page 109)

1

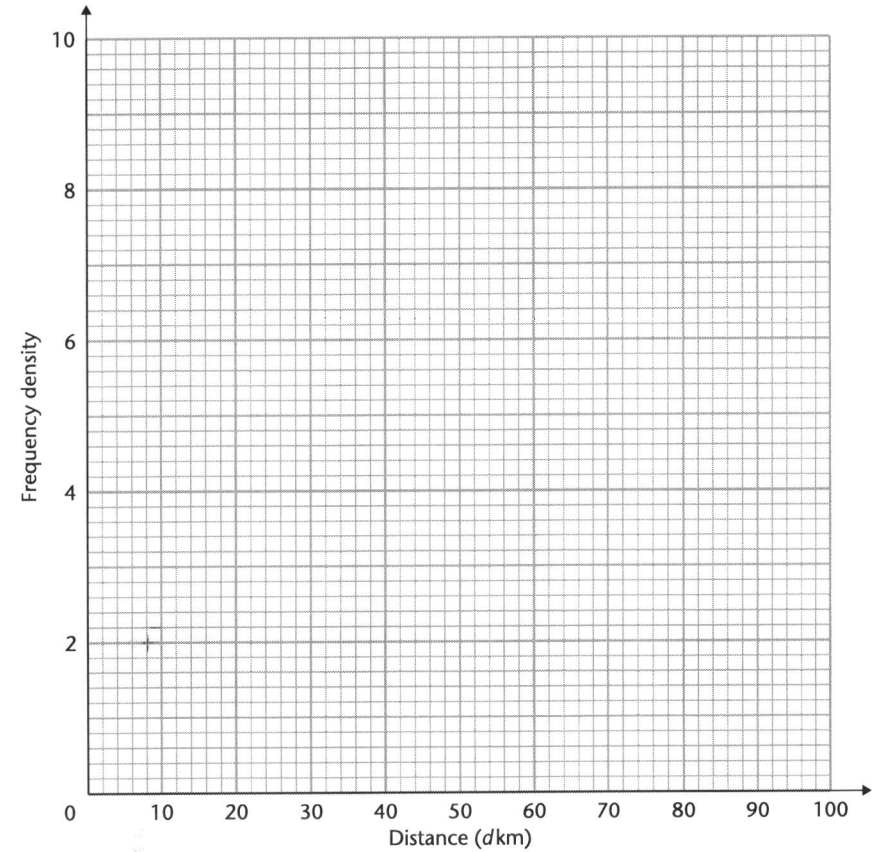

2

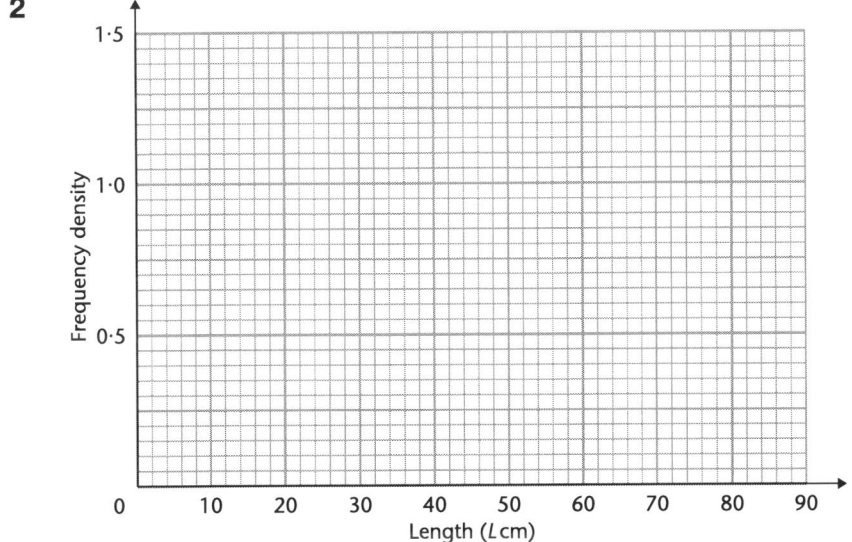

3

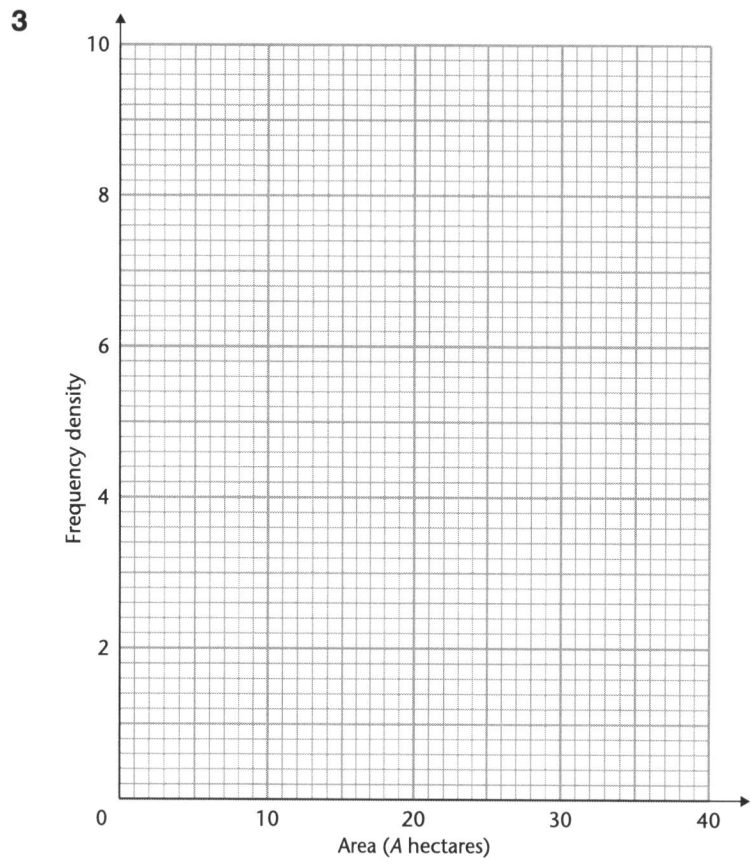

Frequency density vs Area (A hectares)

4

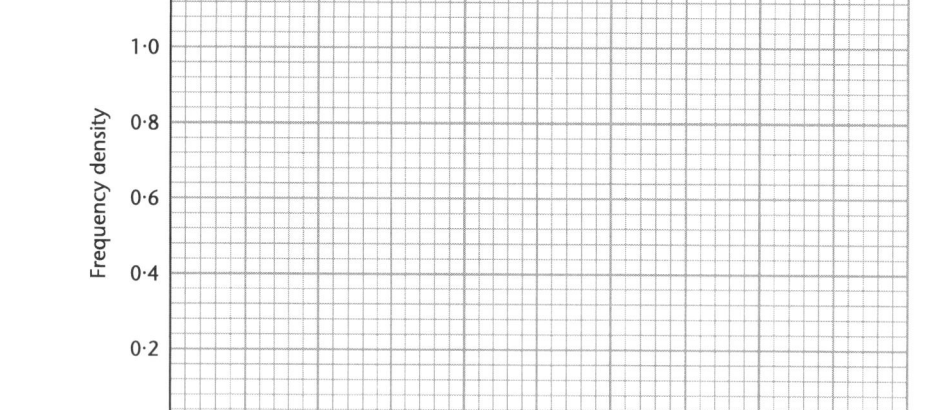

Frequency density vs Age (A years)

Exercise 11.1 (page 109)

1

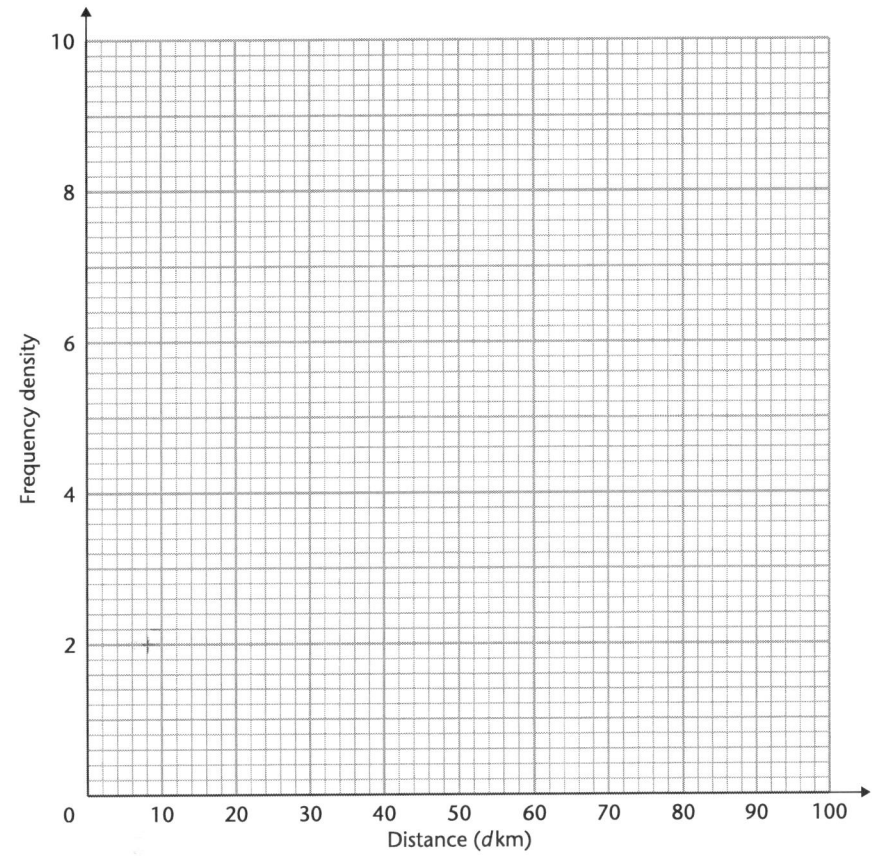

2

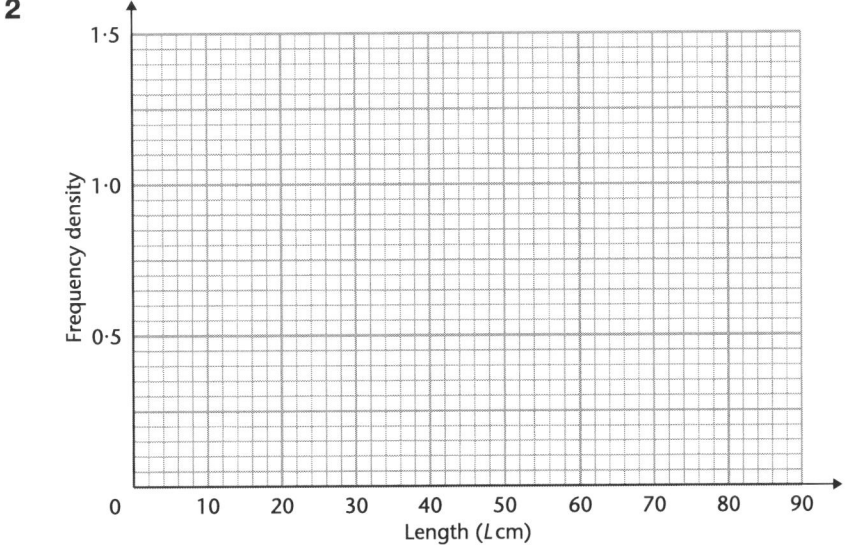

STAGE
9

Speed-up sheets

3

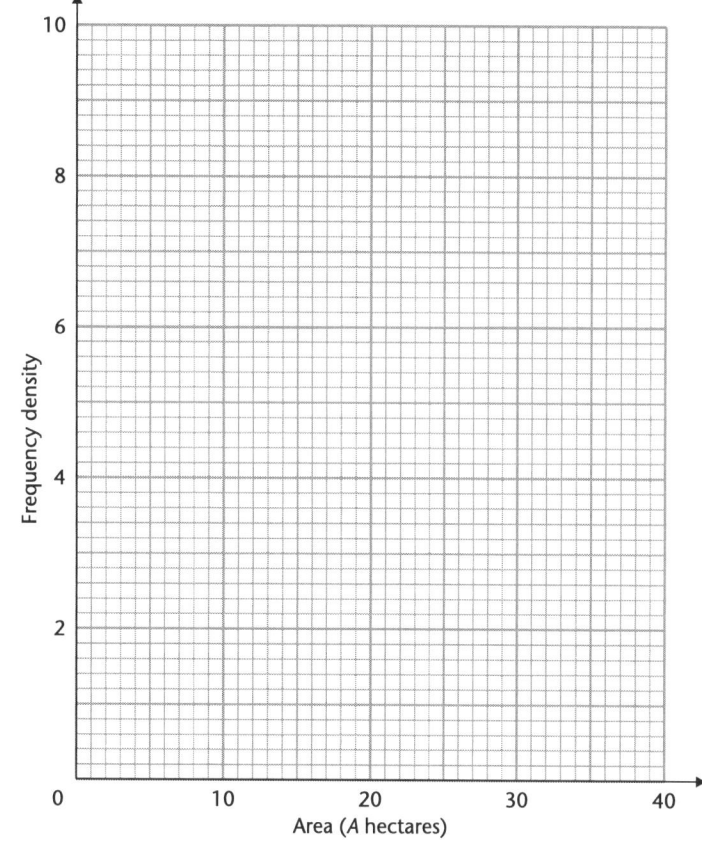

4

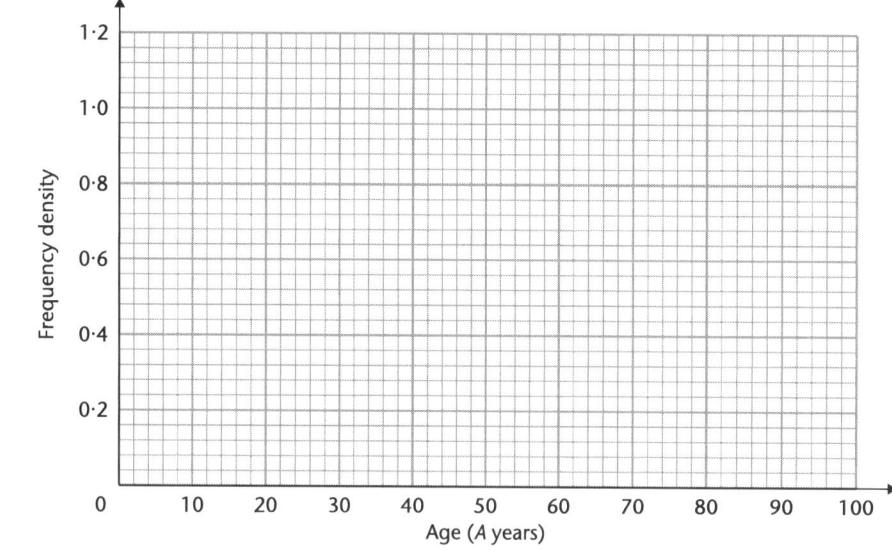

5

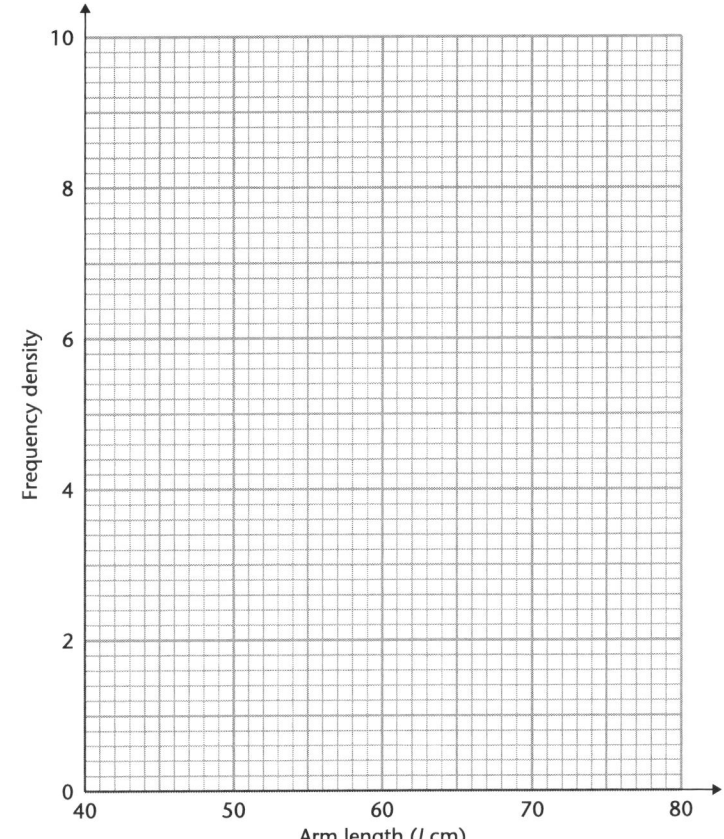

STAGE
9

Speed-up sheets

6

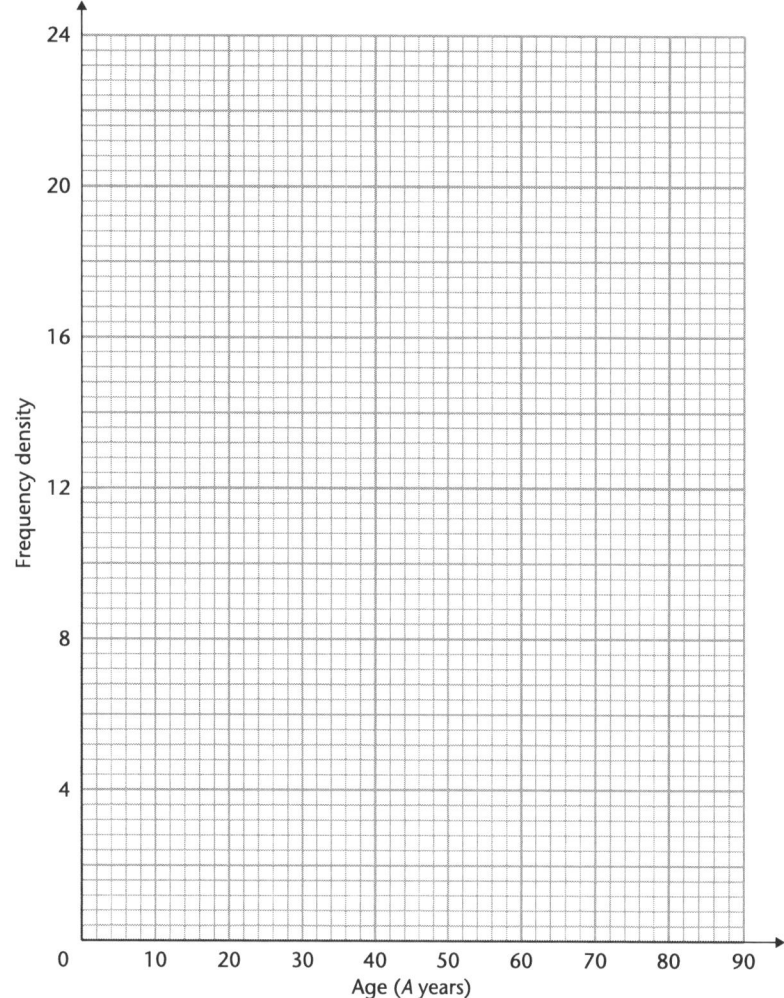

7

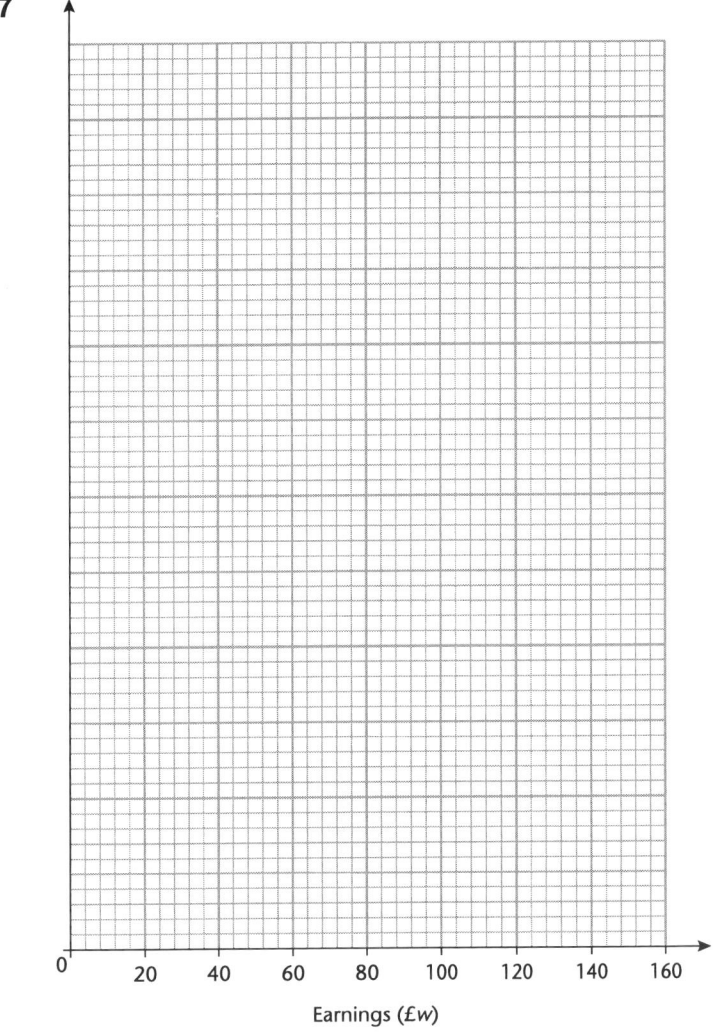

Earnings (£w)

8

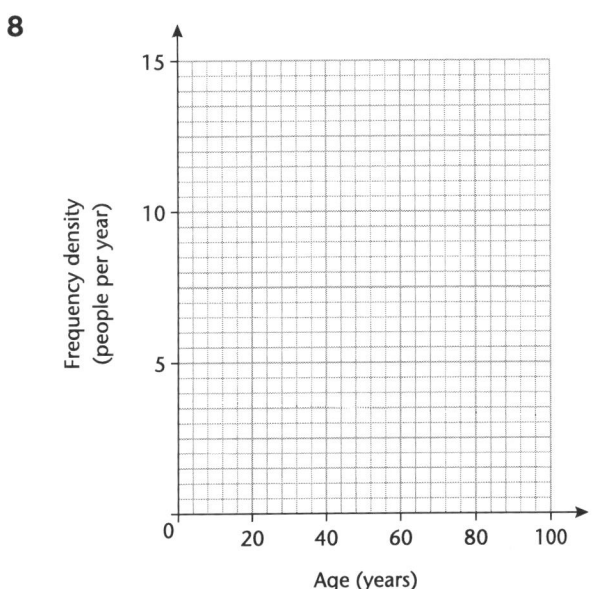

Age (years)

SPEED-UP SHEET 13.1

Exercise 13.1 (page 130)

6

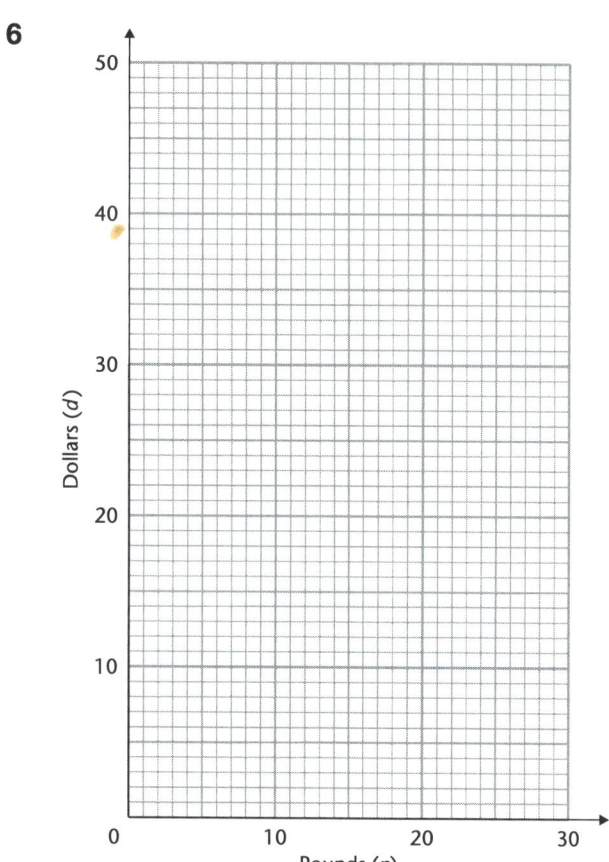

Activity 1 (page 132)

a) (i)

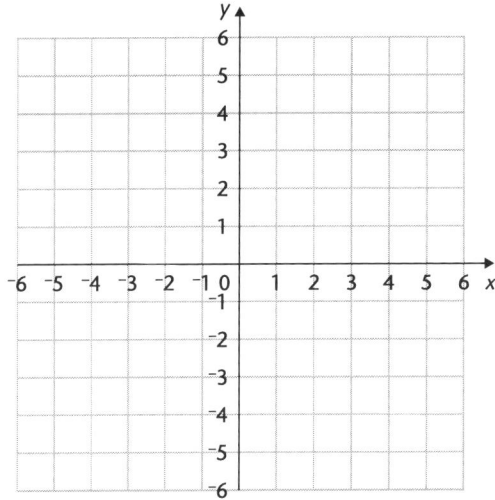

(ii)

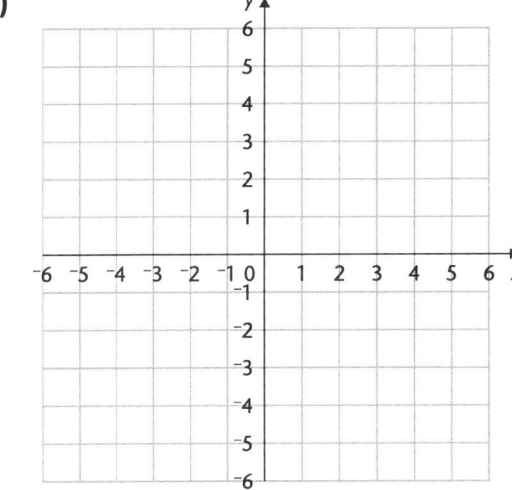

(iii)

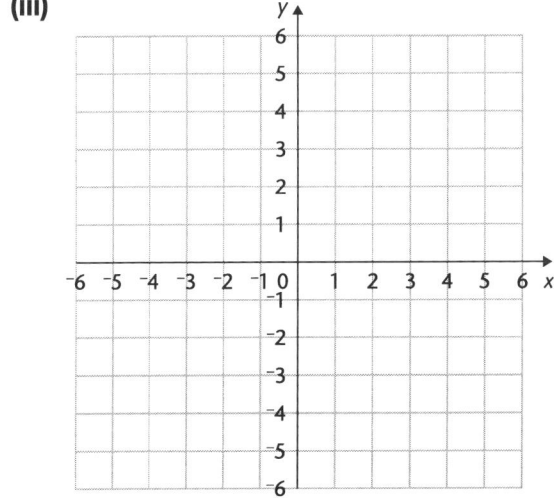

(iv)

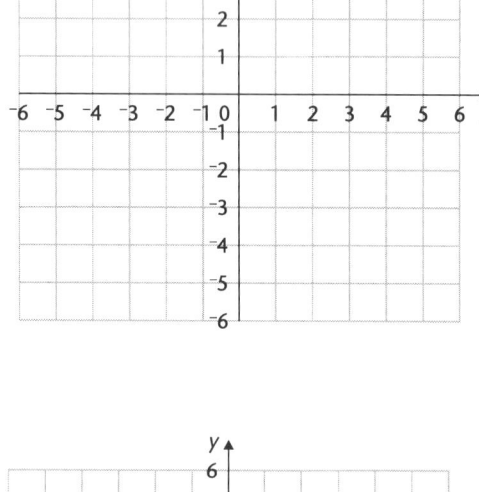

(v)

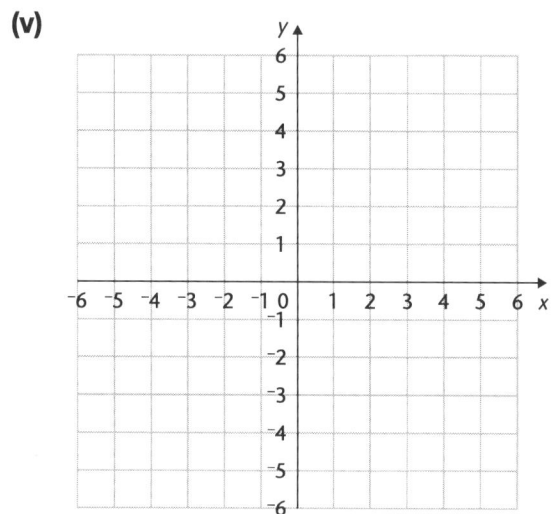

Graduated Assessment for OCR GCSE Mathematics © Hodder Murray 2007

Exercise 13.2 (page 133)

1

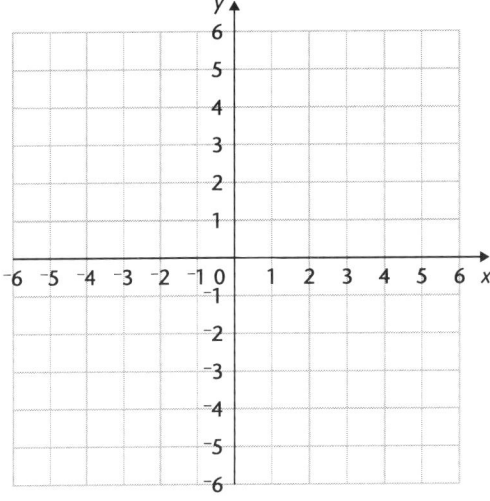

2

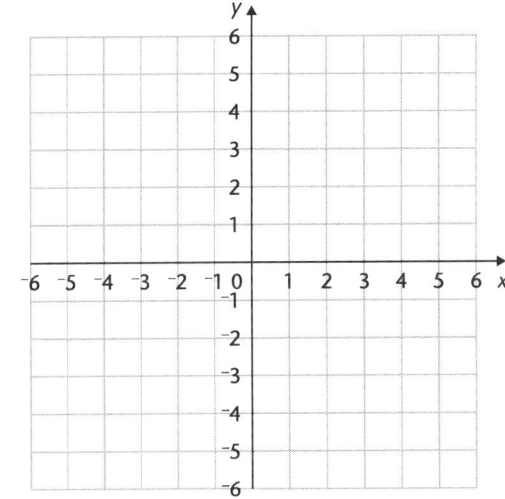

3

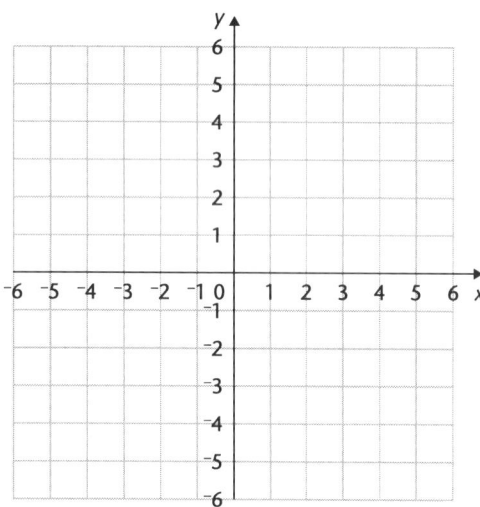

4

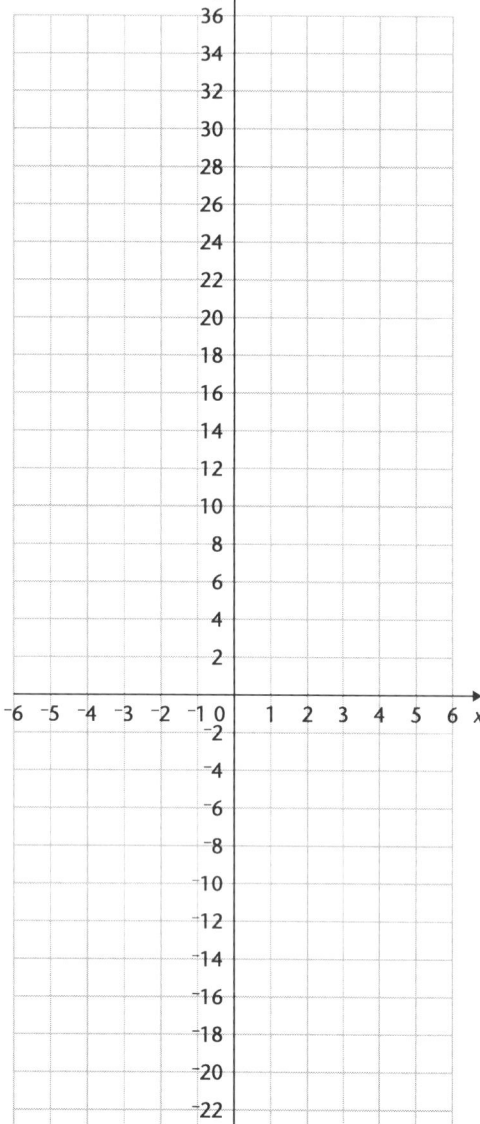

Graduated Assessment for OCR GCSE Mathematics © Hodder Murray 2007

STAGE
9

Revision exercise D1 (page 149)

1

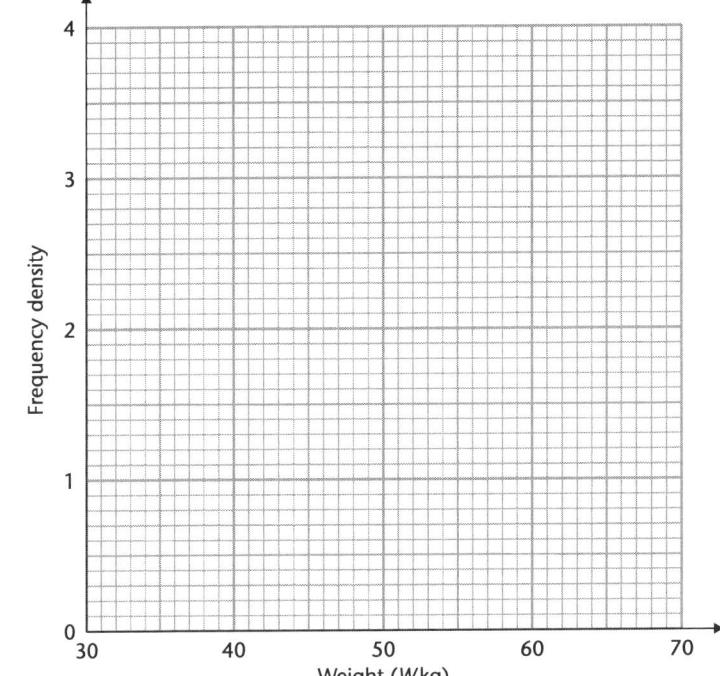

3

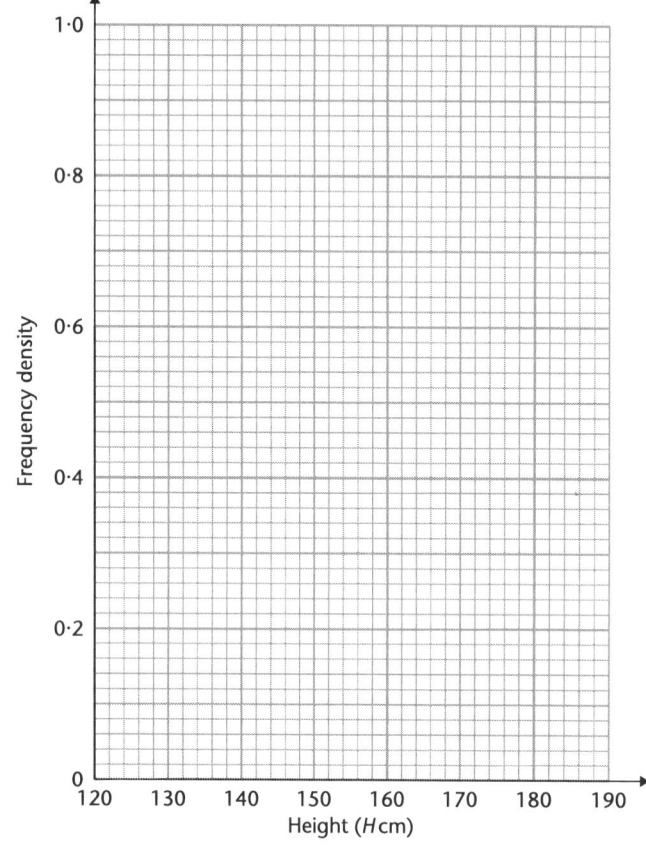

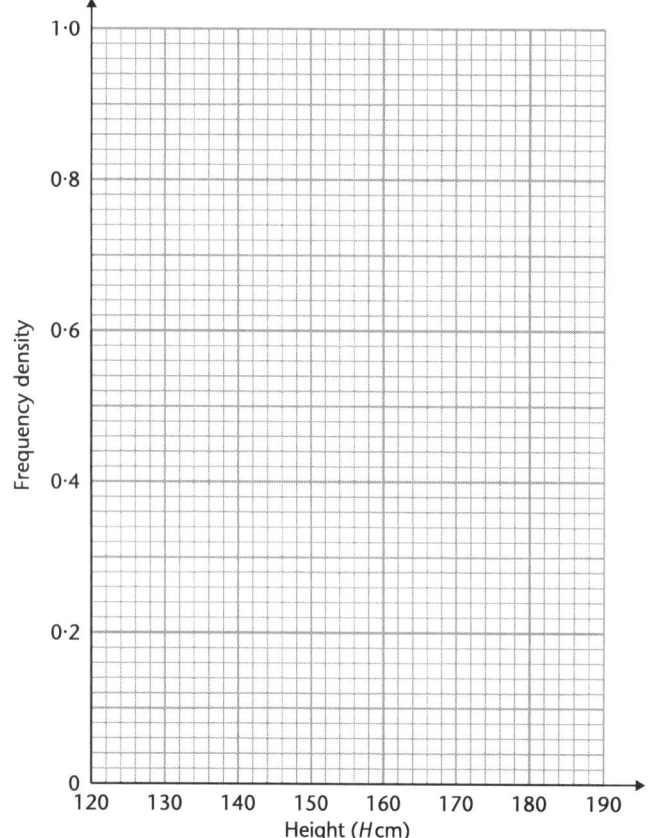

STAGE
9

Stage 9 Answers

1 Checking answers

Exercise 1.1 (page 3)

1. **a)** $500 \times 2 = 1000, 1085{\cdot}64$
   b) $10 \div 5 = 2, 2{\cdot}9$
   c) $4 \times 7 = 28, 27{\cdot}99$
   d) $900 \div 100 = 9, 8{\cdot}90$
   e) $7 \times 10 = 70, 69{\cdot}84$
   f) $100 \div 5 = 20, 20{\cdot}63$
   g) $300 \times 0{\cdot}7 = 210, 212{\cdot}84$
   h) $4 \div 0{\cdot}5 = 8, 8{\cdot}78$

2. **a)** $200 \div 500 = 0{\cdot}4, 0{\cdot}31$
   b) $40 \times (6 + 6) = 480, 445{\cdot}28$
   c) $\sqrt{40\,000} = 200, 203{\cdot}22$
   d) $0{\cdot}1 \div 0{\cdot}1 = 1, 1{\cdot}23$
   e) $(6 - 4) \times 40 = 80, 70{\cdot}98$
   f) $30^2 = 900, 1169{\cdot}64$
   g) $1000 \times 20 = 20\,000, 17\,496$
   h) $0{\cdot}4 \times 0{\cdot}4 = 0{\cdot}16, 0{\cdot}15$
   i) $^-20 \div 5 = {}^-4, {}^-3{\cdot}77$

3. **a)** $\dfrac{3 \times 4}{6} = 2, 1{\cdot}53$
   b) $\dfrac{0{\cdot}2 \times 90}{100 \div 10} = 1{\cdot}8, 1{\cdot}82$
   c) $4 \times \sqrt{400} = 80, 74{\cdot}56$
   d) $\dfrac{500}{20} + \dfrac{500}{10} = 75, 73{\cdot}30$
   e) $\dfrac{30 \times 3}{9} = 10, 9{\cdot}85$
   f) $100 \times 25 = 2500, 2321{\cdot}77$
   g) $\dfrac{50 + 50}{10 + 10} = 5, 4{\cdot}90$
   h) $\dfrac{\sqrt{5 \times 8 \times 0{\cdot}1}}{2^2} = 0{\cdot}5, 0{\cdot}72$

4. **a)** $1 \times 10^{10}, 9{\cdot}21 \times 10^9$
   b) $1{\cdot}4 \times 10^5, 1{\cdot}08 \times 10^5$
   c) $1{\cdot}6 \times 10^2, 1{\cdot}67 \times 10^2$
   d) $3 \times 10^{-3}, 2{\cdot}99 \times 10^{-3}$
   e) $2 \times 10^6, 2{\cdot}18 \times 10^6$
   f) $8 \times 10^8, 6{\cdot}45 \times 10^8$
   g) $9{\cdot}8 \times 10^4, 1{\cdot}11 \times 10^5$
   h) $2{\cdot}5 \times 10^5, 4{\cdot}34 \times 10^5$
   i) $2{\cdot}7 \times 10^{11}, 3{\cdot}02 \times 10^{11}$
   j) $8 \times 10^2, 8{\cdot}74 \times 10^2$
   k) $2 \times 10^1, 1{\cdot}8 \times 10^1$
   l) $4 \times 10^{-4}, 2{\cdot}44 \times 10^{-4}$
   m) $4 \times 10^{-1}, 5{\cdot}38 \times 10^{-1}$
   n) $1{\cdot}6 \times 10^{-10}, 2{\cdot}34 \times 10^{-10}$
   o) $4 \times 10^5, 4{\cdot}62 \times 10^5$
   p) $1{\cdot}2 \times 10^{-11}, 1{\cdot}32 \times 10^{-11}$

5. **a)** $7 \times £9 = £63$
   b) $30 \times £15 = £450$
   c) $3 \times £6 + 3 \times £2 = £24$

6. **a)** **(i)** $40^3 = 64\,000$
   (ii) $20 \times 0{\cdot}2 = 4$
   (iii) $\sqrt{7^2 - 4^2} = \sqrt{49 - 16} = \sqrt{33} \approx 6$
   (iv) $\dfrac{10 + 30}{0{\cdot}08} = \dfrac{40}{0{\cdot}08} = \dfrac{4000}{8} = 500$
   b) **(i)** $60\,236{\cdot}29$ (to 2 d.p.)
   (ii) $3{\cdot}496$
   (iii) $5{\cdot}93$ (to 2 d.p.)
   (iv) 516 (to nearest whole number)

2 Algebraic manipulation

Exercise 2.1 (page 7)

1. $x^2 - x - 6$
2. $x^2 + 3x + 2$
3. $x^2 + 14x + 45$
4. $x^2 + 3x - 18$
5. $x^2 - 2x + 1$
6. $x^2 - 8x + 16$
7. $x^2 - x - 30$
8. $x^2 + 7x + 10$
9. $x^2 - 49$
10. $x^2 - 64$
11. $10x^2 - 22x + 4$
12. $6x^2 - 19x + 10$
13. $12x^2 - 22x - 14$
14. $10x^2 - 3x - 18$
15. $2x^2 + 3xy + y^2$
16. $12x^2 + 7xy + y^2$
17. $3x^2 - 17xy + 20y^2$
18. $2x^2 - 7xy + 6y^2$
19. $12x^2 + xy - 20y^2$
20. $42x^2 + 20xy - 32y^2$
21. $4g^2 - 20gh + 21h^2$
22. $4h^2 - 28hk + 49k^2$
23. $6j^2 - 37jm + 56m^2$
24. $10k^2 + 23kn - 42n^2$
25. $6p^2 - 11pm - 72m^2$
26. $6r^2 - rn - 15n^2$
27. $4q^2 - 4pq - 63p^2$
28. $6r^2 - 37rs + 56s^2$
29. $4s^2 - 20st + 21t^2$
30. $4t^2 - 23t + 15$

Exercise 2.2 (page 10)

1 $12a^5$
2 a^2
3 $2a^2$
4 $12a^4$
5 $9a^6$
6 $8c^3$
7 $6a^5b^3$
8 $6a^5b^7$
9 Cannot simplify
10 $36a^2b^2$
11 $5ab^2$
12 Cannot simplify
13 $3p^3$
14 $4ab^2$
15 $2abc^2$
16 $27b^4$
17 $3t$
18 $5a^2b^3$
19 $2a^3b^3$
20 $2a^3b^2$
21 $\dfrac{4a^2c}{3b^2}$
22 $\dfrac{8x^2y^2}{5z^2}$
23 $\dfrac{a^2b^2c^3}{5}$
24 $\dfrac{6a^6b^3}{5c^3}$
25 $\dfrac{9t^3w^2}{4v^3}$
26 $\dfrac{3e^2g}{f}$
27 $\dfrac{2a^2d^2}{bc^2e}$
28 $\dfrac{z}{tv^2x^3y}$
29 $\dfrac{8a^2c^2}{9b^2d^2}$
30 $\dfrac{5x^2}{2y} + 5x^3$

Exercise 2.3 (page 12)

1 $2(a + 4)$
2 $3(x - 4)$
3 $a(3 + 5a)$
4 $a(4 + 5b)$
5 $2a(b - 3c)$
6 $2a(2b - a)$
7 $5ab(a + 2b)$
8 $a(3b - 2c + 3d)$
9 $x^2y(2y - 3x)$
10 $5(x^2 - 3x + 3)$
11 $3ab(a - 2b)$
12 $ab(4a - 3b)$
13 $2(6x - 3y + 4z)$
14 $3xy(3x - 2y)$
15 $3b(3a + 2b)$

16 $2a^2(7 - 4a)$
17 $2ac(2a - c)$
18 $7(3x^2 - 2y^2)$
19 $5y(3x - 1)$
20 $4xy(3x + 2 - y)$
21 $2a(3a^2 - 2a + 1)$
22 $7st(2s - t)$
23 $3a^2b(1 - 3ab)$
24 $5z(2z^2 - 5z + 1)$
25 $5abc(abc - 2)$
26 $5abc(1 - 3abc)$
27 $a^2b(2 - 3b^2 + 7a^2)$
28 $3abc(a - 2b - 3c)$
29 $a(4bc - 3c^2 + 2ab)$
30 $7a^2b^3c^2(a - 2c)$

Exercise 2.4 (page 14)

1 $(x - 5)(x + 5)$
2 $(x + 2)(x - 2)$
3 $(2a + b)(2a - b)$
4 $(3 - 4y)(3 + 4y)$
5 $(5x - 7y)(5x + 7y)$
6 $(3x - 8)(3x + 8)$
7 $(a - 3b)(a + 3b)$
8 $(1 - 7t)(1 + 7t)$
9 $(10x - 1)(10x + 1)$
10 $(5 - 2x)(5 + 2x)$
11 $(xy - 4a)(xy + 4a)$
12 $(y - 13)(y + 13)$
13 $(11x - 12y)(11x + 12y)$
14 $(9p - 6q)(9p + 6q)$
15 $2(2 - x)(2 + x)$
16 $3(x - 8)(x + 8)$
17 $7(a - 3b)(a + 3b)$
18 $5(3 - 2x)(3 + 2x)$
19 $25(xy - 2)(xy + 2)$
20 $(xyz - 10)(xyz + 10)$
21 $3(x + 2)(x - 2)$
22 $5(x + 3)(x - 3)$
23 $3(x + 6)(x - 6)$
24 $7(x + 7)(x - 7)$
25 $10(x + 20)(x - 20)$
26 $8(x + 5)(x - 5)$

Exercise 2.5 (page 17)

1 $(x + 6)(x + 1)$
2 $(x + 3)(x + 2)$
3 $(x - 4)(x - 2)$
4 $(x - 5)(x - 2)$
5 $2(x + 2)(x + 1)$
6 $(3x + 1)(x + 2)$
7 $(2x + 1)(x + 4)$
8 $(2x + 3)(x + 2)$
9 $3(2x - 1)(x - 2)$
10 $3(x - 2)(x - 2)$
11 $(3x - 2)(x - 3)$
12 $(3x - 10)(x - 1)$
13 $(3x - 5)(x - 2)$
14 $(2x - 5)(2x - 3)$

Graduated Assessment for OCR GCSE Mathematics © Hodder Murray 2007

15 $(2x + 3)(2x + 1)$
16 $(7x + 3)(x + 1)$
17 $(5x - 3)(x - 2)$
18 $(5x - 2)(x - 4)$
19 $(3x - 2)(2x - 5)$
20 $(2x - 3)(4x - 3)$
21 $(3x + 5)(x + 4)$
22 $(2x + 3)(x + 2)$
23 $(3x + 1)(x + 4)$
24 $(5x + 3)(x + 3)$
25 $2(2x + 1)(x + 1)$
26 $(3x + 5)(x + 2)$
27 $(2x + 1)(x + 2)$
28 $(4x + 5)(x + 3)$

Exercise 2.6 (page 19)

1 $(x - 3)(x + 2)$
2 $(x + 3)(x - 6)$
3 $(x + 5)(x - 2)$
4 $(3x - 5)(x + 2)$
5 $(2x - 1)(x + 3)$
6 $2(x - 3)(x + 3)$
7 $(3x + 4)(x - 2)$
8 $(3x + 1)(x - 4)$
9 $(2x - 1)(x + 5)$
10 $(3x - 5)(x + 3)$
11 $5(x - 5)(x + 2)$
12 $(5x - 2)(x + 3)$
13 $(2x + 1)(2x - 3)$
14 $(7x - 4)(x + 2)$
15 $(3x - 7)(x + 2)$
16 $(3x + 4)(x - 5)$
17 $(2x - 7)(x + 3)$
18 $(2x + 1)(x - 8)$
19 $(2x - 7)(3x + 2)$
20 $(6x + 5)(x - 3)$
21 $(2x + 5)(x - 3)$
22 $(3x + 7)(x - 2)$
23 $(5x + 3)(x - 4)$
24 $(3x + 4)(x - 3)$
25 $(4x + 5)(x - 2)$
26 $(2x + 3)(x - 5)$
27 $(4x + 1)(x - 2)$
28 $(3x + 2)(x - 6)$

Exercise 2.7 (page 20)

1 $\dfrac{5ab^2}{2}$

2 $3a^2b$

3 $\dfrac{3x^3y}{20}$

4 $\dfrac{3y^4}{2}$

5 $\dfrac{2}{x - 3}$

6 $\dfrac{x - 4}{2x}$

7 $\dfrac{3x}{x + 3}$

8 $\dfrac{x + 1}{x - 1}$

9 $\dfrac{x - 1}{x + 2}$

10 $\dfrac{3x - 1}{x + 5}$

11 $\dfrac{x - 2}{x + 3}$

12 $\dfrac{x - 4}{3x}$

13 $\dfrac{x - 1}{2x - 5}$

14 $\dfrac{3(x - 2)}{2(x - 3)}$

15 $\dfrac{5(x + 4)}{2x}$

16 $\dfrac{3x - 4}{5(x - 1)}$

17 $\dfrac{3}{x - 2}$

18 $\dfrac{6}{x + 2}$

19 $\dfrac{x - 2}{x - 1}$

20 $\dfrac{x - 4}{x - 5}$

21 $\dfrac{x + 1}{x + 3}$

22 $\dfrac{3(x + 2)}{x + 4}$

23 $\dfrac{3x + 2}{2x - 3}$

24 $\dfrac{2x - 3}{x - 1}$

25 $\dfrac{3x}{2x - 1}$

26 $\dfrac{5(x + 3)}{x - 3}$

27 $x + 2$

28 $\dfrac{x + 2}{2x - 3}$

Exercise 2.8 (page 22)

1 $x = ^-1 \cdot 5$ or $x = 4$
2 $x = \frac{^-2}{3}$ or $x = 4$
3 $x = ^-1 \cdot 5$ or $x = ^-1$
4 $x = ^-1$ or $x = 2 \cdot 5$
5 $x = ^-1$ or $x = \frac{1}{3}$
6 $x = ^-5$ or $x = \frac{^-1}{2}$
7 $x = 1 \cdot 5$ or $x = 5$
8 $x = \frac{^-4}{3}$ or $x = \frac{1}{2}$

9 $x = 5$ or $x = {}^-6$
10 $x = 1$ or $x = 3$
11 $x = {}^-2$ or $x = 3$
12 $x = \frac{-4}{3}$ or $x = 6$
13 $x = \pm 2$
14 $x = \pm 3$
15 $x = \pm 5$
16 $x = \pm 6$
17 $x = \pm 3$
18 $x = \pm 10$

3 Proportion and variation

Exercise 3.1 (page 25)

1 a) $y \propto x$ **b)** $y \propto x$
  c) $t \propto \frac{1}{s}$ **d)** $p \propto w$
  e) $p \propto \frac{1}{n}$ **f)** $d \propto t$
  g) $b \propto \frac{1}{s}$ **h)** $t \propto d$
  i) $c \propto m$ **j)** $p \propto n$

2 a) $y \propto x$ **b)** $y \propto x$
  c) $y \propto \frac{1}{x}$ **d)** $y \propto x$
  e) $y \propto x$ **f)** $y \propto \frac{1}{x}$
  g) $y \propto x$ **h)** $y \propto x$
  i) $y \propto \frac{1}{x}$ **j)** $y \propto x$

Exercise 3.2 (page 28)

1 a) $y = \frac{1}{3}x$

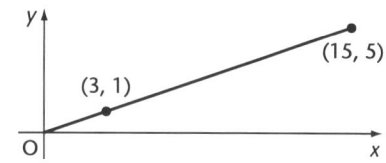

b) $y = 7x$

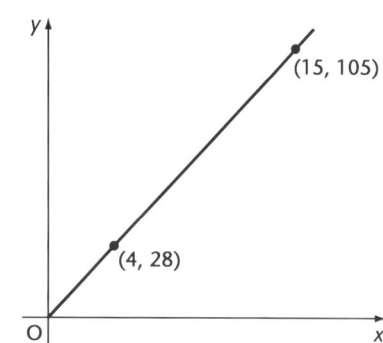

c) $xy = 80$

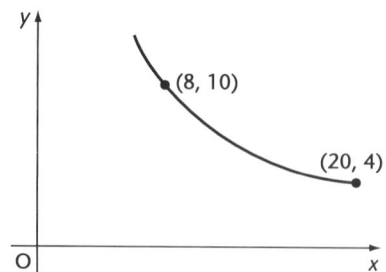

d) $y = 5x$

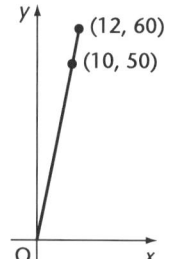

e) $y = \frac{2}{3}x$

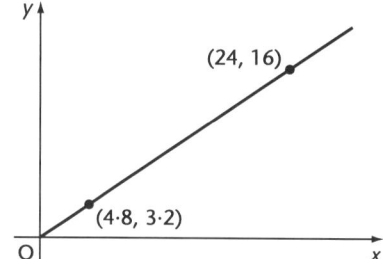

f) $xy = 15$

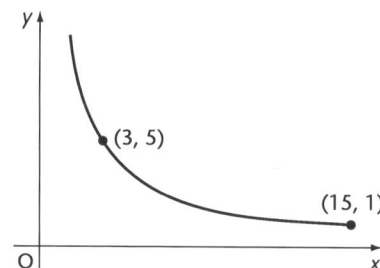

g) $y = \frac{2}{3}x$

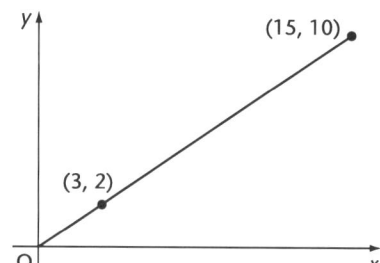

STAGE
9

40   Graduated Assessment for OCR GCSE Mathematics © Hodder Murray 2007

h) $y = \frac{5}{4}x$

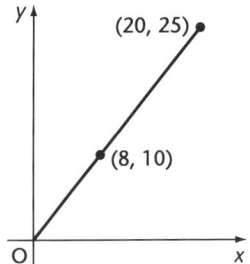

i) $xy = 50$

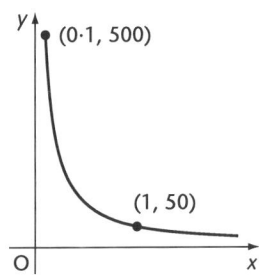

j) $y = \frac{2}{5}x$

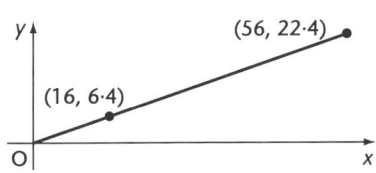

2 a) $I = \frac{V}{6}$ **b)** 10

3 a) $w = \frac{330}{f}$ **b)** 0·5 m

Exercise 3.3 (page 29)

1 12
2 4
3 8
4 1
5 1·25
6 0·625
7 $4\frac{4}{9}$
8 96
9 $\frac{3}{8}$ or 0·375
10 3·9375
11 32
12 32
13 1·75
14 2
15 28
16 12
17 $27\frac{11}{32}$ or 27·343 75
18 18
19 80
20 81

21 a) $y \propto x^2$ **b)** $y \propto \frac{1}{x^2}$

c) $y \propto x^3$ **d)** $y \propto \frac{1}{x}$

e) $y \propto \frac{1}{x^2}$ **f)** $y \propto x^2$

g) $y \propto \frac{1}{x}$ **h)** $y \propto x$

i) $y \propto x^3$ **j)** $y \propto \frac{1}{x^2}$

Exercise 3.4 (page 31)

1 $y = \frac{x^2}{12}$

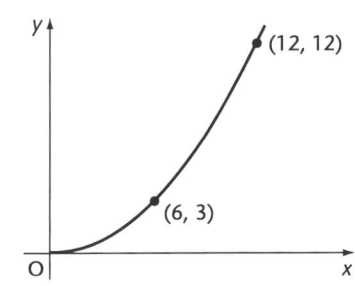

2 $y = \frac{4x^2}{25}$

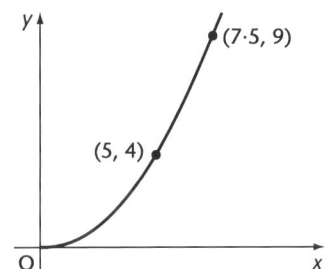

3 $y = \frac{x^3}{27}$

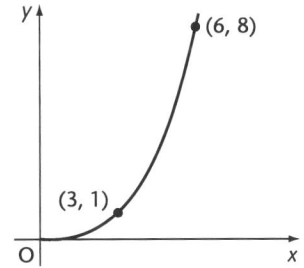

4 $y = \frac{64}{x^2}$

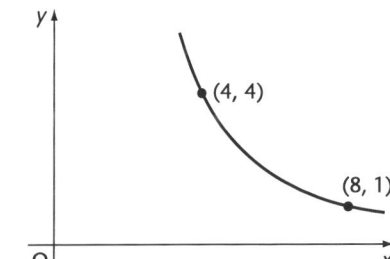

5 $y = \dfrac{5x^2}{36}$

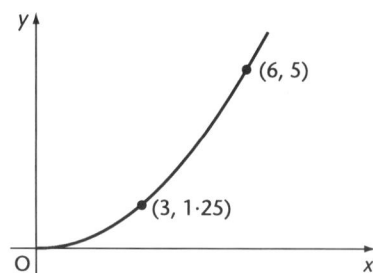

6 $y = \dfrac{90}{x^2}$

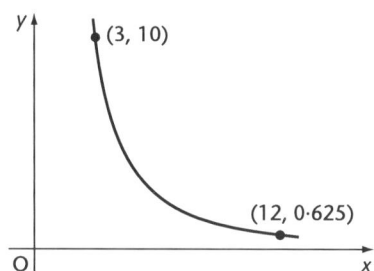

7 $y = \dfrac{160}{x^2}$

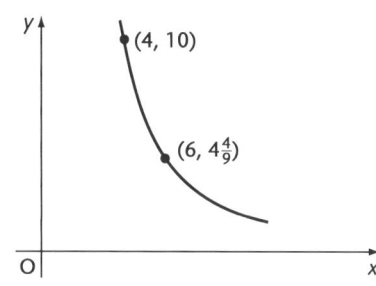

8 $y = \dfrac{12x^3}{125}$

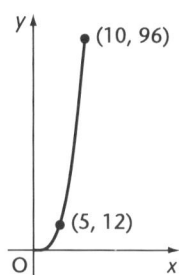

9 $y = 0{\cdot}003x^3$

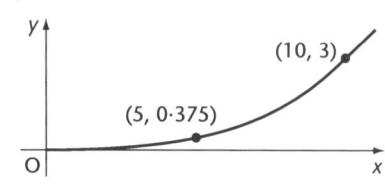

10 $y = \dfrac{7x^2}{64}$

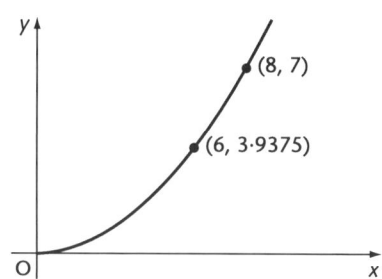

11 $y = \dfrac{4x^3}{125}$

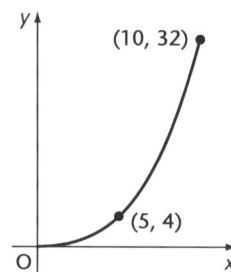

12 $y = \tfrac{1}{2}x^2$

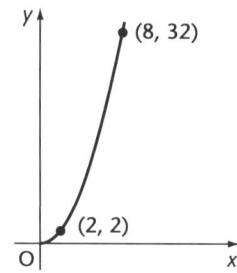

13 $y = \dfrac{343}{x^2}$

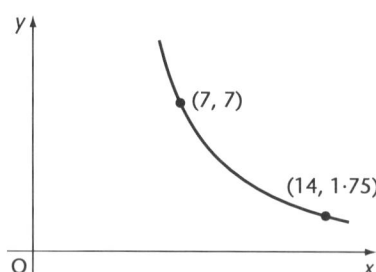

14 $y = \dfrac{1}{x}$

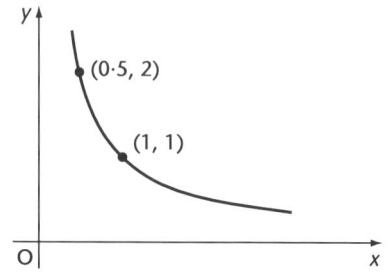

15 $y = \dfrac{8x}{3}$

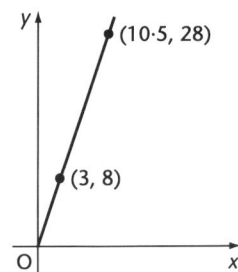

16 $y = \dfrac{3}{x^2}$

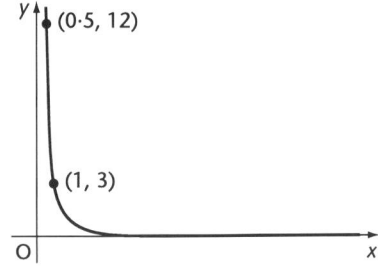

17 $y = \dfrac{7x^3}{864}$

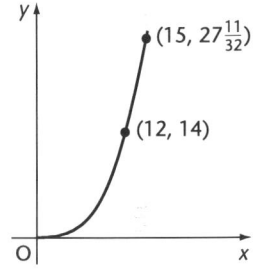

18 $y = \dfrac{8x^2}{169}$

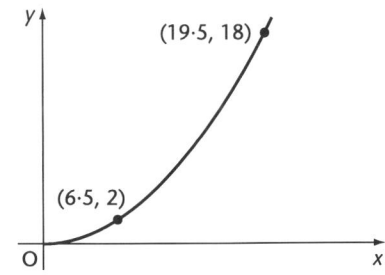

19 $y = \dfrac{125x^2}{9}$

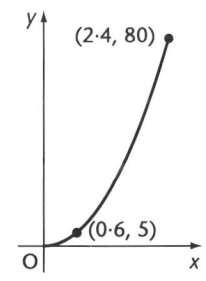

20 $y = \dfrac{81}{x^2}$

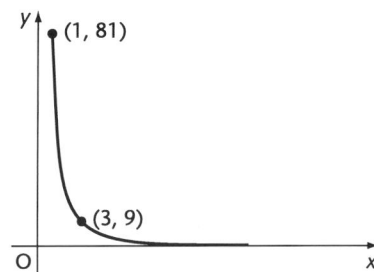

21 a) $y = \dfrac{x^2}{5}$

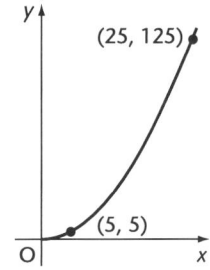

b) $y = \dfrac{125}{x^2}$

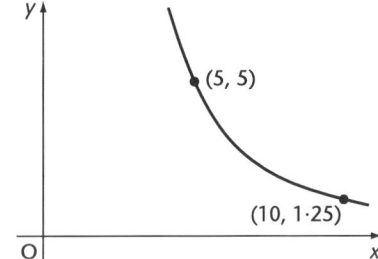

c) $y = 0\cdot04x^3$

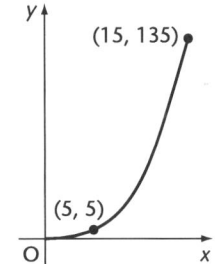

d) $xy = 25$

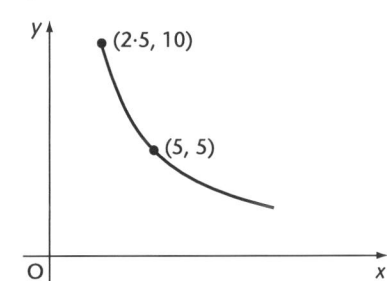

e) $y = \dfrac{288}{x^2}$

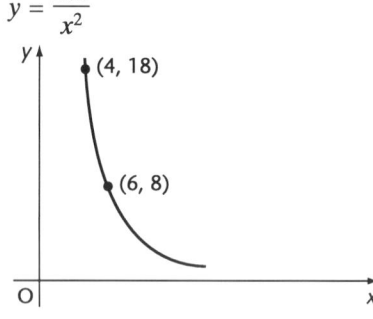

f) $y = \dfrac{7x^2}{4}$

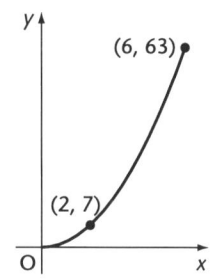

g) $y = \dfrac{1}{x}$

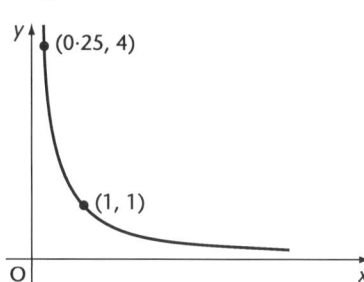

h) $y = \dfrac{11x}{18}$

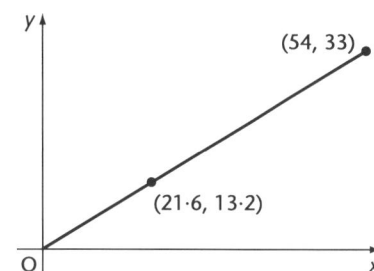

i) $y = \dfrac{15x^3}{4096}$

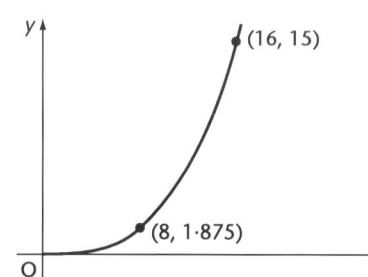

j) $y = \dfrac{2304}{x^2}$

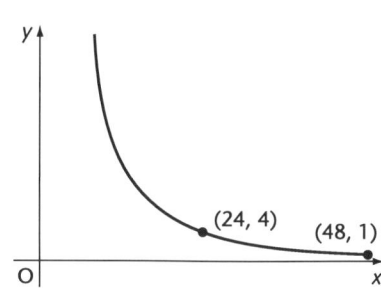

Exercise 3.5 (page 31)

1 a) (i) $y \propto x^2$ **(ii)** $y = x^2$
 b) (i) $y \propto x^2$ **(ii)** $y = 2x^2$
 (iii) 128
 c) (i) $y \propto x^2$ **(ii)** $y = \frac{1}{3}x^2$
 d) (i) $y \propto x^2$ **(ii)** $y = 0{\cdot}4x^2$
 (iii) 490
 e) (i) $y \propto x^2$ **(ii)** $y = 1{\cdot}2x^2$
 f) (i) $y \propto \dfrac{1}{x^2}$ **(ii)** $y = \dfrac{225}{x^2}$
 (iii) 1
 g) (i) $y \propto \dfrac{1}{x^2}$ **(ii)** $y = \dfrac{1250}{x^2}$
 h) (i) $y \propto \dfrac{1}{x^2}$ **(ii)** $y = \dfrac{20}{x^2}$
 (iii) 0·2
 i) (i) $y \propto \dfrac{1}{x^2}$ **(ii)** $y = \dfrac{800}{x^2}$
 j) (i) $y \propto \dfrac{1}{x^2}$ **(ii)** $y = \dfrac{256}{x^2}$
 (iii) 1

2 a) Multiplied by 4
 b) Multiplied by 2·25
3 a) $h = \frac{5}{98}u^2$
 b) 216 m (to the nearest whole number)
4 $1{\cdot}52 \times 10^8\,\text{km}^2$

Revision exercise A1 (page 33)

1 a) $600 \times 80 = 48\,000$, 48 140
 b) $60 \div 4 = 15$, 16·63
 c) $\dfrac{30^2}{0{\cdot}5} = 1800$, 1668·52
2 a) $1{\cdot}458 \times 10^4$
 b) $2{\cdot}385 \times 10^{-1}$
3 a) $x^2 + 8x + 7$
 b) $a^2 + 2a - 15$
 c) $2y^2 - 2y - 4$
 d) $2x^2 - 9x - 5$
 e) $4a^2 - 3ab - b^2$
4 a) $2a^5$ **b)** $5a$
 c) a^5 **d)** $24a^4b^4$
 e) $3xz$ **f)** $4a^2c$

STAGE 9

5 a) $3(a + 2b - 4c)$
  b) $a(2 + 3b)$
  c) $ab(a - 3b)$
  d) $2xy(x - 3)$
  e) $7ab(c + 2a)$
  f) $3(3a^2 + b^2 - 2c^2)$
  g) $5(pq - 2)$
  h) $2a(1 - 2a + 3a^2)$
  i) $50ac(2b - 1)$

6 a) $(x - 7)(x + 7)$
  b) $(x - 1)(x + 1)$
  c) No factors
  d) $(x - y)(x + y)$
  e) $(11 - b)(11 + b)$

7 a) $(x - 9)(x - 7)$
  b) $2(x + 3)(x - 7)$
  c) $(x - 2)(3x - 2)$
  d) $(x + 3)(2x - 5)$
  e) $3(x - 4)(x + 4)$
  f) $(2x + 3)(x - 7)$
  g) $3(2x + 1)(x - 5)$
  h) $(5x - 6)(x - 3)$
  i) $(2x + 1)(4x - 5)$
  j) $(3x + 2)(2x - 5)$

8 a) $\dfrac{x + 3}{4}$ **b)** $\dfrac{x + 1}{x - 2}$

  c) $\dfrac{2x}{x + 3}$ **d)** $\dfrac{x - 2}{2x - 1}$

  e) $\dfrac{x + 3}{x - 5}$ **f)** $\dfrac{3(x + 1)}{2x - 1}$

9 a) $x = \frac{-4}{3}$ or $x = 3$
  b) $x = 3 \cdot 5$ or $x = {}^{-}4$
  c) $x = \pm 8$

10 a) $0 \cdot 06$ units   **b)** $F = \dfrac{54}{d^2}$

11 $25, 125, 625, 1, 0 \cdot 2$ ✗

12 a) (i) $y \propto \dfrac{1}{x}$ **(ii)** $xy = 50$

  b) (i) $y \propto x$ **(ii)** $y = 2x$

  c) (i) $y \propto \dfrac{1}{x}$ **(ii)** $xy = 0 \cdot 2$

  d) (i) $y \propto x^2$ **(ii)** $y = 0 \cdot 025x^2$

4 Indices

Exercise 4.1 (page 38)

1 a) $n^{\frac{1}{3}}$ **b)** n^{-3}
  c) $n^{\frac{2}{5}}$ **d)** $n^{\frac{1}{4}}$
  e) n^4 **f)** $n^{\frac{5}{3}}$

2 a) $\frac{1}{4}$ **b)** 2
  c) 1 **d)** $\frac{1}{16}$
  e) 8

3 a) 2 **b)** $\frac{1}{8}$
  c) 16 **d)** 64
  e) 8

4 a) $\frac{1}{9}$ **b)** 3
  c) 1 **d)** $\frac{1}{81}$
  e) 27

5 a) 3 **b)** 81
  c) $\frac{1}{27}$ **d)** 3
  e) 1

6 a) 8 **b)** $\frac{1}{4}$
  c) 1 **d)** 16
  e) 32

7 a) 4 **b)** $\frac{1}{2}$
  c) 1 **d)** 64
  e) 128

8 a) 12 **b)** 64
  c) $\frac{1}{3}$ **d)** 27

9 a) 125 **b)** 6
  c) 100 **d)** $\frac{343}{3} = 114\frac{1}{3}$

10 a) 9 **b)** 16
  c) $\frac{16}{3} = 5\frac{1}{3}$ **d)** 15

11 a) 1000 **b)** $\frac{2}{25}$
  c) $113\frac{3}{4}$ **d)** 0

12 a) $1 \cdot 925\,414 = 1 \cdot 9254$ (to 5 s.f.)
  b) $21 \cdot 717\,639 = 21 \cdot 718$ (to 5 s.f.)
  c) $1 \cdot 045\,910 = 1 \cdot 0459$ (to 5 s.f.)
  d) $0 \cdot 003\,538\,869 = 0 \cdot 003\,538\,9$ (to 5 s.f.)

13 a) $111 \cdot 5664 = 111 \cdot 57$ (to 5 s.f.)
  b) $0 \cdot 020\,596\,29 = 0 \cdot 020\,596$ (to 5 s.f.)
  c) $1 \cdot 072\,135 = 1 \cdot 0721$ (to 5 s.f.)
  d) $0 \cdot 040\,580\,84 = 0 \cdot 040\,581$ (to 5 s.f.)

14 a) 31
  b) $1 \cdot 009\,998\,068 = 1 \cdot 0100$ (to 5 s.f.)
  c) $1 \cdot 544\,857 = 1 \cdot 5449$ (to 5 s.f.)
  d) $2 \cdot 107\,773 = 2 \cdot 1078$ (to 5 s.f.)

15 a) 11
  b) $6 \cdot 760\,315 = 6 \cdot 7603$ (to 5 s.f.)
  c) $1 \cdot 276\,518 = 1 \cdot 2765$ (to 5 s.f.)
  d) 16

16 a) $106 \cdot 1208 = 106 \cdot 12$ (to 5 s.f.)
  b) $13 \cdot 427\,845 = 13 \cdot 428$ (to 5 s.f.)
  c) $25 \cdot 304\,39 = 25 \cdot 304$ (to 5 s.f.)

17 a) 145
  b) $7 \cdot 483\,282 = 7 \cdot 4833$ (to 5 s.f.)
  c) $1 \cdot 422\,970 = 1 \cdot 4230$ (to 5 s.f.)

18 a) $3 \cdot 7711$
  b) $2 \cdot 167\,981 = 2 \cdot 1680$ (to 5 s.f.)
  c) 0

19 a) $3547 \cdot 171 = 3547 \cdot 2$ (to 5 s.f.)
  b) $4 \cdot 732\,081 = 4 \cdot 7321$ (to 5 s.f.)
  c) $299 \cdot 532\,070\,5 = 299 \cdot 53$ (to 5 s.f.)

20 a) 18 **b)** 243
  c) $\frac{1}{5}$ **d)** 18
  e) 16 **f)** 72
  g) $\frac{36}{5} = 7\frac{1}{5}$ **h)** 26

Exercise 4.2 (page 41)

1 a) 3^3 b) 3^{-1}
  c) $3^{\frac{3}{2}}$ d) 3^6
  e) 3^2 f) 3^{11n}
2 a) 2^5 b) 2^2
  c) 2^3 d) 2^{-2}
  e) 2^{3n} f) 2^{3n-8}
3 a) 5^4 b) 5^2
  c) 5^{-1} d) $5^{-\frac{5}{2}}$
  e) Cannot simplify f) 5^{7n}
4 a) 7^3 b) $5^{\frac{1}{3}}$
  c) 2^{-2} d) 1
  e) Cannot simplify f) 3^{2n}
5 a) $2^3 \times 3$ b) $2^6 \times 3^2$
  c) $2^{\frac{1}{3}} \times 3^{\frac{2}{3}}$ d) $2^2 \times 3^{-2}$ or $\frac{2^2}{3^2}$
  e) $3^3 \times 2^{-1}$ or $\frac{3^3}{2}$ f) $2^{4n} \times 3^{2n}$
6 a) $2^2 \times 3^2$ b) $2^5 \times 3$
  c) $2^2 \times 3 \times 5$ d) $2^3 \times 7^2$
  e) 3×5^2 f) $2^4 \times 3^2$
  g) $2^2 \times 3 \times 5^2$ h) $2^2 \times 3^4$
7 a) $3^3 \times 5^3$ b) 2
  c) $2^{3n} \times 5^n$ d) $2^{6n} \times 5^{4n}$
8 a) 5^{-1} b) 5^3
  c) 5^4 d) 5^5
  e) Cannot simplify f) 5^{7n}

5 Rearranging formulae

Exercise 5.1 (page 44)

1 $t = \dfrac{s}{a+2b}$

2 $b = \dfrac{s}{a-c}$

3 $t = \dfrac{Pb}{b-a}$

4 $a = \dfrac{1}{s-b}$

5 $x = \dfrac{s+bs}{b+2a}$

6 $y = \dfrac{7-3a}{3-b}$

7 $t = \dfrac{ab}{1-bs}$

8 $c = \dfrac{1-ab}{a}$

9 $a = \dfrac{b+2}{2+2b}$ or $\dfrac{b+2}{2(1+b)}$

10 $a = \dfrac{b^2}{1-2b}$

11 $d = \dfrac{b+c-a}{a-b}$

12 $b = \dfrac{100a}{m+100}$

13 $c = \sqrt{(a-b)}$

14 $P = \dfrac{100A}{100+RT}$

15 $p = \dfrac{a}{1-a}$

16 $x = \dfrac{1-a-b}{a+b}$

17 $x = \dfrac{b-a}{2a-b}$

18 $L = \dfrac{T^2 g}{4\pi^2}$

19 $r = \sqrt{\dfrac{s+1}{2}}$

20 $v = \dfrac{us}{u-s}$

21 $t = \dfrac{2s}{2u+a}$

22 $x = \dfrac{b+a}{3a-2b}$

23 $v = \dfrac{fu}{u-f}$

24 $x = \sqrt{\dfrac{y+4}{3}}$

25 $h = \sqrt{\left(\dfrac{A}{\pi r}\right)^2 - r^2}$

26 $u = \sqrt{v^2 - 2as}$

27 $r = \sqrt{\dfrac{3V}{\pi h}}$

28 $t = \sqrt{\dfrac{30-2s}{a}}$

6 Arcs, sectors and volumes

Exercise 6.1 (page 49)

1 a) 3·56 cm b) 13·6 cm
  c) 27·0 cm d) 12·4 cm
  e) 8·41 cm f) 5·91 cm
  g) 15·7 cm h) 26·5 cm
  i) 4·00 cm j) 12·8 cm
2 a) 9·08 cm² b) 25·1 cm²
  c) 139 cm² d) 59·5 cm²
  e) 18·1 cm² f) 13·9 cm²
  g) 21·2 cm² h) 127 cm²
  i) 8·00 cm² j) 41·1 cm²

3 a) 25·7 cm   b) 25·9 cm
  c) 25·8 cm   d) 26·3 cm
  e) 51·3 cm   f) 11·7 cm
4 a) 43°   b) 100°
  c) 185°   d) 58°
  e) 148°   f) 57°
  g) 179°   h) 203°
  i) 159°   j) 62°
  k) 188°   l) 257°
5 a) 4·9 cm   b) 3·4 cm
  c) 4·8 cm   d) 5·6 cm
  e) 3·7 cm   f) 2·5 cm
  g) 13·4 cm   h) 5·55 cm
  i) 4·62 cm
6 a) 6·59 cm   b) 1·51 cm
  c) 1·81 m
7 Blue area = 626 mm² (nearest mm²)
  Black length = 167 mm (nearest mm)
8 a) 344 m² (nearest m²)
  b) 77 m (nearest m)
9 186·7°, 62·6 cm²
10 208 cm

Exercise 6.2 (page 54)

1 a) 18 cm³   b) 54 cm³
  c) 70 m³   d) 50 cm³
  e) 179 cm³   f) 30 cm³
2 a) 103 cm³   b) 314 cm³
  c) 51·5 cm³   d) 154 cm³
  e) 1010 cm³   f) 181 cm³
3 a) 524 cm³   b) 998 cm³
  c) 33·5 mm³   d) 113 cm³
  e) 435 cm³   f) 1988 mm³
4 a) 6·57 cm³   b) 121 cm³
5 12 cm
6 a) 3·6 cm   b) 3·1 cm
  c) 6·2 cm
7 a) 556 cm³   b) 2310 cm³
  c) 4190 cm³
8 a) 6·6 cm   b) 12·4 cm
9 117 ml
10 7·96 cm
11 88
12 69
13 69 100 mm³
14 12 cm
15 3·17 cm

7 Upper and lower bounds

Exercise 7.1 (page 60)

1 a) 61·2 seconds   b) 24·51 seconds
  c) 12·4 m   d) 1·747 kg
  e) 185 mm   f) 12·738 kg
  g) 148·3 cm   h) 105·86 seconds
2 a) 61 seconds   b) 24·49 seconds
  c) 12·38 m   d) 1·745 kg
  e) 183 mm   f) 12·736 kg
  g) 148·1 cm   h) 105·84 seconds

3 a) 704 g   b) 6·7 cm
  c) 790 g   d) 4·4 seconds
  e) 0·16 second   f) 76 m
  g) 9000 m   h) 0·138 g
4 a) 702 g   b) 6·5 cm
  c) 770 g   d) 4·2 seconds
  e) 0·14 second   f) 74 m
  g) 8800 m   h) 0·136 g
5 50·8 cm
6 a) 29·20 seconds   b) 1·06 seconds
7 UB = 26·5 cm, LB = 23·5 cm
8 UB = 3 kg, LB = 2·8 kg
9 UB = 1 cm, LB = 0 cm
10 a) UB = 13·8, LB = 13·6
   b) UB = 3·6, LB = 3·4

Exercise 7.2 (page 62)

1 No
2 75 mm
3 a) 16·3625 m²   b) 20·507 175 m²
  c) 56·910 875 m²   d) 40·1625 m²
4 a) 15·5625 m²   b) 20·415 675 m²
  c) 56·759 175 m²   d) 38·8825 m²
5 a) 1141·7575 cm   b) 12·746 25 m
  c) 146·625 km   d) 11·707 275 m
6 a) 1131·2875 cm   b) 12·510 25 m
  c) 138·425 km   d) 11·630 375 m
7 a) 5·08 cm/second   b) 1·25 m/second
  c) 10·5 m/second
8 a) 5·61 cm/second   b) 1·28 m/second
  c) 10·7 m/second
9 0·022 88 kg/cm³
10 a) 11·7 cm   b) 10·6 cm
   c) 20·4 cm
11 a) 2·54 cm   b) 5·17 cm
   c) 3·66 m
12 UB = 844, LB = 830 people/square mile
13 UB = 536·25 cm², LB = 490·25 cm²
14 UB = 7·88 m/second, LB = 7·74 m/second
15 a) UB = 87·95 cm³, LB = 81·37 cm³
   b) UB = 1702 g, LB = 1566 g
16 UB = 84·5, LB = 20·2
17 UB = 0·405, LB = 0·384

Revision exercise B1 (page 64)

1 a) n^{-1} b) $m^{\frac{1}{3}}$
  c) $n^{-\frac{1}{2}}$
2 a) $\frac{1}{4}$ b) 1
  c) 5   d) 16
  e) 4   f) $\frac{1}{25}$
  g) 2   h) 32
  i) $1\frac{1}{2}$ j) 144
3 a) 625   b) 80
  c) 9   d) 8
  e) 42   f) 500
  g) 4   h) $2\frac{1}{12}$

4 a) 2·9242 **b)** 0·498 42
 c) 4096 **d)** 0·000 256 74
5 a) 9 **b)** 1·6765
 c) 3·9842 **d)** 512
6 a) 2^7 **b)** 3^6
 c) $7^{\frac{2}{3}}$ **d)** 3^6
 e) 2^{6n+5}
7 a) $2^3 \times 5$ **b)** $2 \times 3^2 \times 5$
 c) $2^3 \times 17$ **d)** $2^2 \times 3 \times 7^2$
8 a) $3^2 \times 5^2$ **b)** $2 \times 5^{\frac{1}{3}}$
 c) $2^4 \times 7^{\frac{10}{3}}$ **d)** $2^3 \times 3^{\frac{5}{2}}$
9 a) $a = \dfrac{2b - 5c}{8}$ **b)** $a = 2b$

10 a) $p = \dfrac{q}{7}$ **b)** $p = \dfrac{2}{2 - t}$

 c) $p = \dfrac{qs}{s + q}$ **d)** $p = \dfrac{Tq}{2 + q}$

 e) $p = \dfrac{1 + 2a}{4a - 1}$ **f)** $p = \pm \sqrt{2b - 4a}$

11 a) 8·51 cm **b)** 27·7 cm²
12 a) 66·0° **b)** 144°
13 a) 6·21 cm **b)** 4·84 cm
14 7240 cm³
15 136 m³
16 UB = 535 g, LB = 505 g
17 a) 31·2 cm **b)** 8·2 cm
18 a) UB of space 1000·5 mm
  UB of each unit 500·5 mm,
  so UB of two units 1001 mm
  (or other valid explanation).
 b) 1·5 mm
19 UB = 101 cm³, LB = 94·2 cm³
20 a) 7·38 m/second **b)** 7·35 m/second
21 978·75 m
22 602 people/km²

8 Similarity and enlargement

Exercise 8.1 (page 68)

1 a) 4 **b)** 9
 c) 25 **d)** 16
 e) 36 **f)** 100
2 a) 1000 **b)** 64
 c) 125 **d)** 8
 e) 27 **f)** 512
3 a) 4 **b)** 6
 c) 8 **d)** 10
4 25 cm²
5 a) 72·5 cm² **b)** 18·1 m²
6 1 litre
7 25·9 cl
8 360 cm²
9 5·2 m²
10 1:50

11 1:1·59
12 12·6 cm
13 12·96 m²
14 27:64:125
15 12·5 cm
16 a) 2 m **b)** 3 m²
 c) 18 750 cm³
17 2·48 m
18 a) 3·6 m **b)** 5·76 litres
19 a) 15 **b)** 225
 c) 4·52 m³
20 15·1 cm
21 0·0226 m² or 226 cm²
22 3·30:2·08:3·78 or 1·59:1:1·82
23 77·44 cm²

Exercise 8.2 (page 71)

1 a) (1, 8) **b)** ⁻3
2

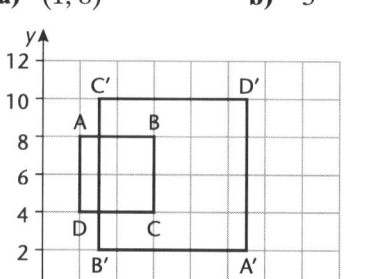

3 a) (0, 3) **b)** ⁻2
4 a) (2, 3) **b)** ⁻0·5

9 Probability

Exercise 9.1 (page 77)

1 $\frac{5}{10} = \frac{1}{2}$
2 0·65
3 $\frac{8}{52} = \frac{2}{13}$
4 $\frac{16}{100} = \frac{4}{25}$
5 0·24
6 $\frac{3}{6} = \frac{1}{2}; \frac{1}{4}$
7 $\frac{144}{2704} = \frac{9}{169}$
8 0·56
9 0·85
10 0·04
11 Don't add up to 1
12 a) 0·75 **b)** 0·175
13 a) $\frac{1}{13}$ **b)** $\frac{1}{169}$
14 a) 0·6, 0·4 **b)** 0·24
15 $\frac{1}{144}$
16 $\frac{1}{27}$
17 a) $\frac{15}{23}$ **b)** $\frac{18}{23}$
18 a) 0·85 **b)** 0·65

STAGE
9

19 a) 0·53   **b)** 0·168
   c) 0·412   **d)** 0·83
   e) 0·56

20 The events are not independent: $\frac{3}{5} + \frac{1}{2} > 1$.

21 $\frac{1}{4}$

22 $\frac{1}{144}$

Exercise 9.2 (page 81)

1 $\frac{5}{8}$

2 $\frac{2}{499}$

3 $\frac{12}{51} = \frac{4}{17}$

4 $\frac{79}{198}$

5 a)

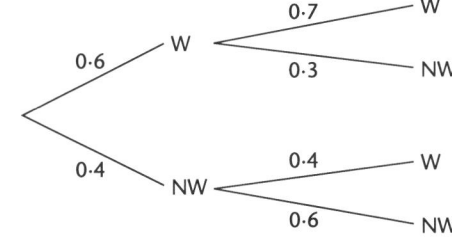

   b) 0·34   **c)** 0·76

6 a)

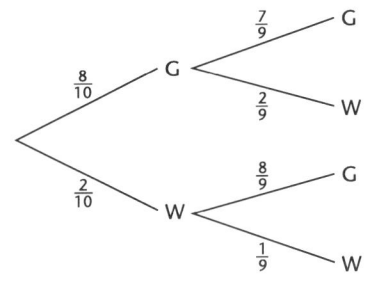

   b) $\frac{2}{90} = \frac{1}{45}$

   c) $\frac{32}{90} = \frac{16}{45}$

7 $\frac{2}{999\,000}$

8 $\frac{68}{110} = \frac{34}{55}$

9 a) $\frac{20}{90} = \frac{2}{9}$ **b)** $\frac{28}{90} = \frac{14}{45}$

10 0·5

11 0·76

12 0·86

13 $\frac{50}{56} = \frac{25}{28}$

14 $\frac{38}{132} = \frac{19}{66}$

15 $\frac{24}{132\,600} = \frac{1}{5525}$

16 a) $\frac{12}{72} = \frac{1}{6}$ **b)** $\frac{20}{72} = \frac{5}{18}$

   c) $\frac{40}{72} = \frac{5}{9}$

17 a) $\frac{156}{2652} = \frac{1}{17}$ **b)** $\frac{12}{2652} = \frac{1}{221}$

   c) $\frac{32}{2652} = \frac{8}{663}$

18 a)

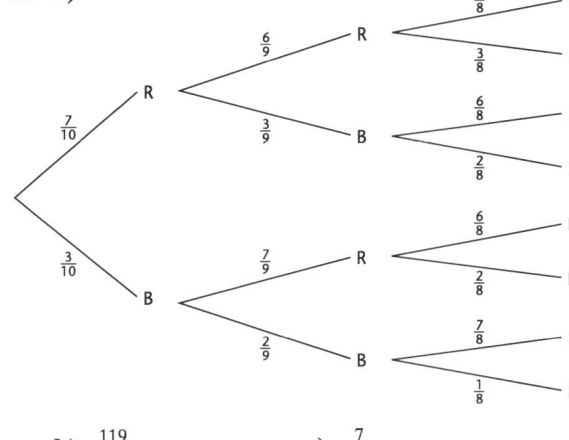

   b) $\frac{119}{120}$ **c)** $\frac{7}{40}$

10 Working in two and three dimensions

Exercise 10.1 (page 87)

1 a) 5   **b)** 7·21
   c) 7·28   **d)** 9·43
   e) 6·40   **f)** 7·62

2 a) 6·40   **b)** 7·07
   c) 4·47   **d)** 6·71
   e) 6·32   **f)** 13

Exercise 10.2 (page 91)

1 a) (i) 41·6°   **(ii)** 15 cm
    (iii) 17 cm   **(iv)** 28·1°

   b) A(0, 0, 0), B(0, 9, 0), C(12, 9, 0),
      D(12, 0, 0), E(0, 0, 8), F(0, 9, 8),
      G(12, 9, 8), H(12, 0, 8)

2 a) 11·3 cm   **b)** 27·9°
   c) 12·8 cm   **d)** 51·3°

3 a) AC = 102·5 m, BC = 64·0 m
   b) 120·9 m   **c)** 328°

4 a) 8·47 cm   **b)** 45·1°

5 a) 45°   **b)** 73·4°
   c) 11·3 cm   **d)** 5·66 cm
   e) 12·8 cm   **f)** 66·2°

6 a) 7·1 cm   **b)** 3·9 cm
   c) No. $QU^2 \neq QR^2 + RU^2$

7 a) 33·8 cm   **b)** 94·3 cm
   c) 21°   **d)** 61 300 cm^3

8 a) (i) 17 cm   **(ii)** 13·1 cm
    (iii) 69·7°
   b) (i) 10·8 cm   **(ii)** 68·2°

9 a) 5 cm   **b)** 26·8°

10 a) 21·2 cm   **b)** 16·8 cm
   c) 16·8 cm

11 a) (i) 5·3 m   **(ii)** 2·1 m
    (iii) 19·3°   **(iv)** 4·5 m

   b) A(0, 0, 0), B(4, 0, 0), C(4, 2, 0), D(0, 2, 0),
      R(0, 0, 2·1), S(4, 0, 2·1), T(4, 2, 2·8),
      H(0, 2, 2·8)

12 Yes. CE = 11·96 m and AC = 87·3 m

Also AC = $\sqrt{80^2 + 35^2}$ = 87·3 m

13 a) 11·0 cm   **b)** 35·5°

14 a) 10·9 m   **b)** 68·9°

     c) 11·7 m

Exercise 10.3 (page 95)

1 a) 10·63 units to 2 d.p.

     b) 10·82 units to 2 d.p.

     c) 15·33 units to 2 d.p.

     d) 5·48 units to 2 d.p.

     e) 12·37 units to 2 d.p.

2 6 or ⁻6

Exercise 10.4 (page 98)

1 a)

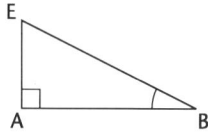

     b)

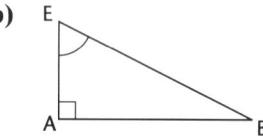

     c)

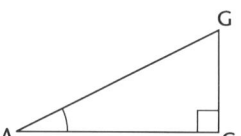

     d)

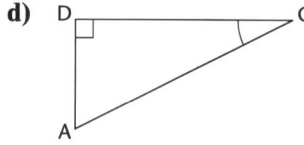

2 a) 32°   **b)** 58°

     c) 26·6°   **d)** 32·5°

3 a) 33·7°   **b)** 36·5°

4 a)

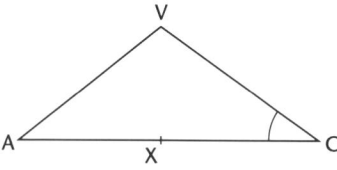

     b)

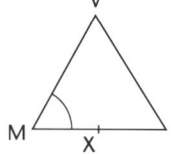

5 a) 65·4°   **b)** 74·6°

6 67·4°

7 a) 23·4°   **b)** 49·3°

8 62·1°

9 a) 11·6 cm   **b)** 11·7 cm

10 a) 22·9 cm

     b) (i) 12·6°   **(ii)** 60·8°

11 35·3°; all cubes are similar

Revision exercise C1 (page 100)

1 a) 180 cm²   **b)** 6 cm; 95 cm²

     c) 614·4 ml; 1200 ml

2 800 ml

3 7·42 cm

4

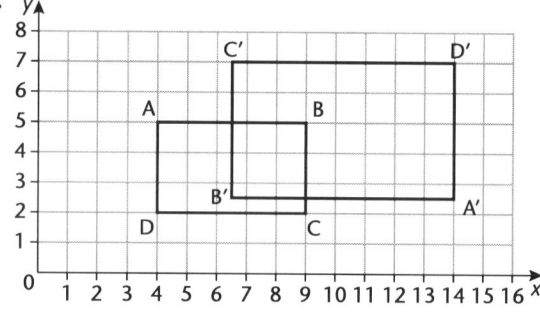

A′(14, 2·5), B′(6·5, 2·5), C′(6·5, 7), D′(14, 7)

5 a) (0, 4)   **b)** ⁻2

6 0·85

7 0·13

8 a) 0·09   **b)** 0·42

9 a) 0·25   **b)** 0·75

     c) 0·08

10 a) 0·01   **b)** 0·18

11 0·57

12 $\frac{11}{60}$

13 a) 5·39 units   **b)** 9·22 units

14 a) 56·8°   **b)** 8·96 m

15 11·0 cm

16 a) 5·5 units   **b)** 12·4 units

     c) 13 units

17 a) Angle BHC   **b)** Angle AGB

18 a) Angle YBV = 51·1°

     b) Angle CAD = 35·0°

     c) 12·8 m

19 a) 149°   **b)** 5·85 km

     c) 4·9°

Graduated Assessment for OCR GCSE Mathematics © Hodder Murray 2007

11 Histograms

Exercise 11.1 (page 106)

1

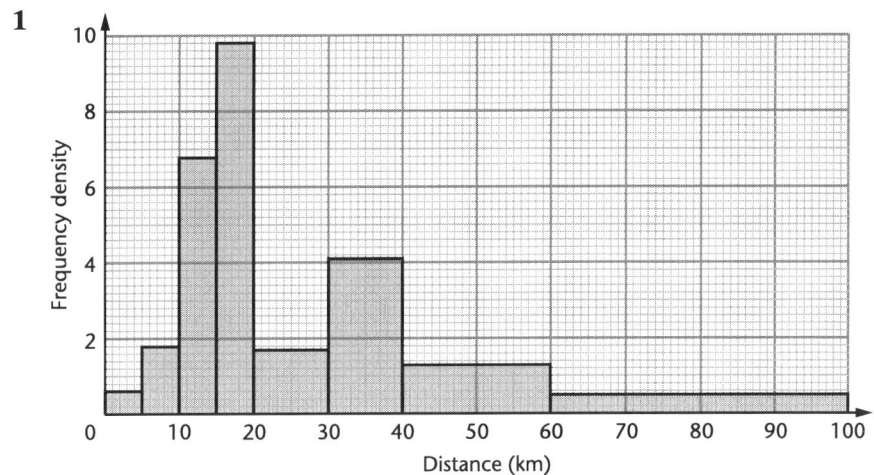

2

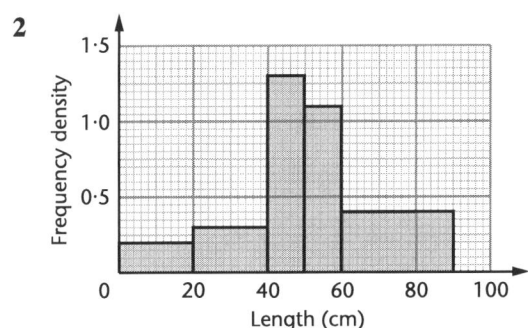

3

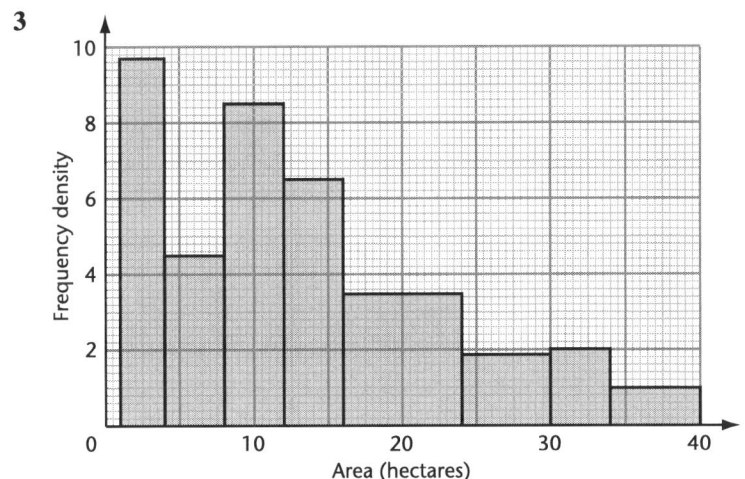

4

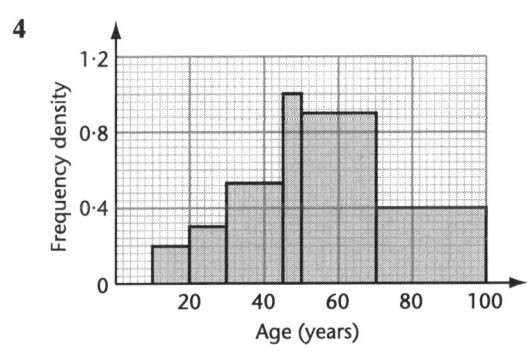

5

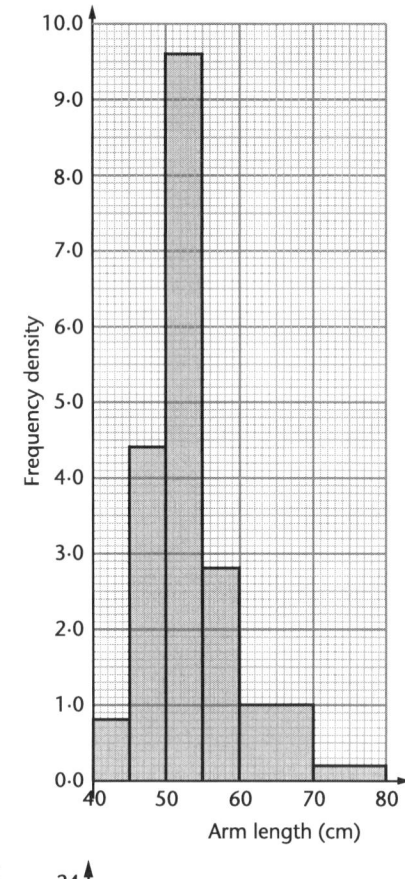

6

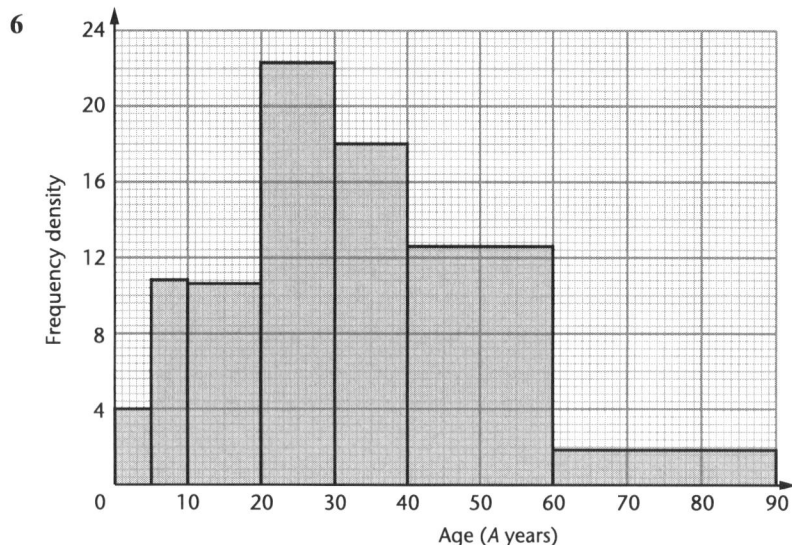

7

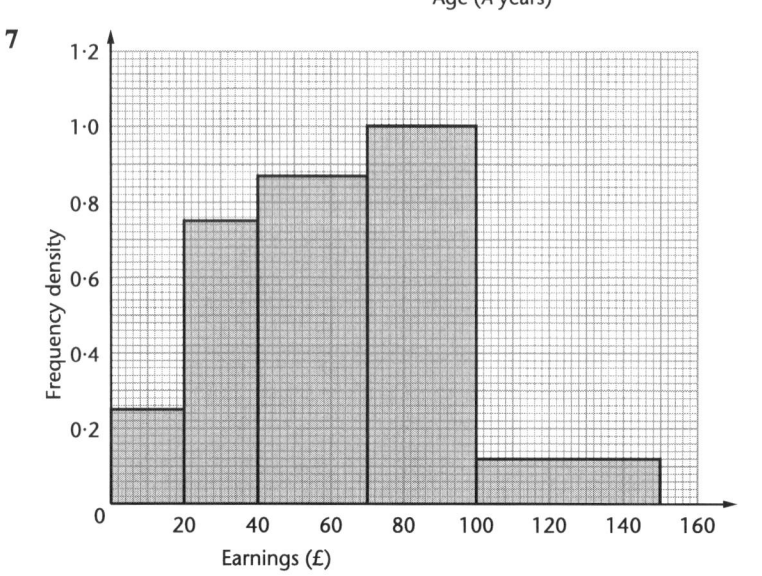

Graduated Assessment for OCR GCSE Mathematics © Hodder Murray 2007

8 a) You are aged 19 up until your 20th birthday, which is 10 years after your 10th birthday.

b)

Age (years)	Under 10	10–19	20–29	30–49	50–89
Frequency	24	46	81	252	288
Frequency density	2·4	4·6	8·1	12·6	7·2

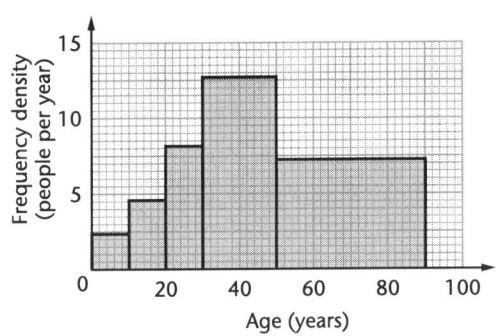

Exercise 11.2 (page 109)

1 a) 130 **b)** 36·2 years
2 a) 31 **b)** 43·4 kg
3 a) 490 **b)** 25·5 minutes
4 a) 134 **b)** 10·4 km
5 a) 100 **b)** 81·2 kg to 1 d.p.

12 Circle properties

Other reasons may be possible in some cases.

Exercise 12.1 (page 115)

1 $a = 140°$ (angle at the circumference $= \frac{1}{2}$ angle at the centre)
2 $b = 45°$ (angle at the centre $= 2 \times$ angle at the circumference)
    $c = 45°$ (isosceles triangle)
3 $d = 100°$ (isosceles triangle)
    $e = 50°$ (angle at the circumference $= \frac{1}{2}$ angle at the centre)
4 $f = 60°$ (sum of angles on a straight line)
    $g = 120°$ (angle at the centre $= 2 \times$ angle at the circumference)
5 $h = 25°$ (angle at circumference $= 45°$, isosceles triangle)
6 $i = (180° – 136°) \div 2 = 22°$ (isosceles triangle)
7 $j = 80°$ (isosceles triangle)
    $k = 40°$ (angle at the centre $= 2 \times$ angle at the circumference)
8 $l = 45°$ (angle at the centre $= 2 \times$ angle at the circumference)
    $m = 135°$ (sum of angles on straight line)
9 $n = 110°$ (angle at the centre $= 2 \times$ angle at the circumference)
    $o = 35°$ (isosceles triangle)

10 $p = 66°$ (angle in a semicircle and angle sum of a triangle)
11 $q = 49°$ (angle at the circumference $= \frac{1}{2}$ angle at the centre)
    $r = 41°$ (angle in a semicircle and angle sum of a triangle)
12 $s = 42°$ (angle at the circumference $= \frac{1}{2}$ angle at the centre)
13 $t = 30°$ (angle in a semicircle and angle sum of a triangle)
14 $u = 120°$ (angle at the circumference $= \frac{1}{2}$ angle at the centre)

Exercise 12.2 (page 118)

1 $a = 50°$ (angles in the same segment)
    $b = 100°$ (angles at centre and circumference)
2 $c = 110°$ (angles around a point and angles at centre and circumference)
    $d = 110°$ (angles in the same segment)
3 $e = 45°$ (angles at centre and circumference)
    $f = 45°$ (angles in the same segment)
4 $g = 45°$ (angles at centre and circumference)
    $h = 45°$ (angles in the same segment)
    $i = 45°$ (isosceles triangle)
5 $j = 80°$ (angles at centre and circumference)
    $k = 100°$ (opposite angles of cyclic quadrilateral)
6 $l = 140°$ (angles at centre and circumference)
    $m = 110°$ (opposite angles of cyclic quadrilateral)
7 Angle at centre $= 120°$ (isosceles triangle)
    Reflex angle at centre $= 240°$
    Therefore $n = 120°$ (angles at centre and circumference)
8 $o = 20°$ (angles in the same segment)
    $p = 40°$ (angles in the same segment)
    $q = 60°$ (external angle of a triangle)
9 $r = 104°$ (opposite angles of cyclic quadrilateral)
    $s = 85°$ (opposite angles of cyclic quadrilateral)
10 $t = 50°$ (angles in the same segment)
    $u = 56°$ (angle sum of a triangle and angles in the same segment)
    $v = 34°$ (opposite angles of cyclic quadrilateral)
11 $w = 45°$ (opposite angles of cyclic quadrilateral)
    $x = 60°$ (opposite angles of cyclic quadrilateral)
12 $y = 40°$ (angles in the same segment)
    $z = 80°$ (angles at centre and circumference)
    $a = 50°$ (angle sum of an isosceles triangle)
13 $b = 95°$ (angles on a straight line and opposite angles of cyclic quadrilateral)
    $c = 126°$ (opposite angles of cyclic quadrilateral and angles on a straight line)

14 $d = 30°$ (angle in a semicircle, angle sum of a triangle and angles in same segment)

15 $e = 90°$ (angle in a semicircle)
$f = 40°$ (angles in the same segment)
$g = 32°$ (angle sum of a triangle)

Exercise 12.3 (page 120)

1 AC is a diameter
Angle ABC = 90° (angle in semicircle)
Angle OAB = 40° (angle sum of triangle ABC)
Angle $a = 40°$ (angle sum of triangle ABO)
Angle $b = 40°$ (angles subtended by same arc)
Angle DOC = 100° (vertically opposite angles)
Angle $c = 50°$ (angles at centre and circumference)

2 $d = 60°$ (angles in same segment)
Angle CEA = 120° (angles on straight line)
$e = 180° - (40° + 120°) = 20°$ (angle sum of triangle)
$f = 20°$ (angles in same segment)

3 $g = 69°$ (angles at centre and circumference)
$h = 69° \div 2 = 34·5°$ (exterior angle of triangle ADC = sum of interior opposite angles)

4 $i = 90°$ (angle in semicircle)
$j = 80°$ (angle at centre)
$k = 50°$ (angles in isosceles triangle)

5 $l = 50°$ (angles in same segment)
Angle at centre = 100°
$m = 40°$ (isosceles triangle)

6 Angle DAB = angle DBA = 70° (isosceles triangle DAB)
Angle OAB = angle OBA = 30° (isosceles triangle OAB)
Therefore $n = 40°$

7 Angle AXC = 180° - 125° = 55° (angles on straight line)
$o = 55°$ (angles in same segment)
Angle ABC = 125° (opposite angles of cyclic quadrilateral)
$p = (180° - 125°) \div 2 = 27·5°$ (base angles of isosceles triangle)

8 $q = 90°$ (angle in semicircle)
$r = (180° - 90° - 60°) = 30°$ (angle sum of triangle)
$s = r = 30°$ (subtended by same arc)
$t = 180° - 90° - (20° + 30°) = 40°$ (angle sum of a triangle)

9 $u = 98°$ (opposite angles of cyclic quadrilateral)
Angle DEX = 60° (angle sum of a triangle)
$v = 120°$ (opposite angles of cyclic quadrilateral)

10 Angle PQR = 90° (angle in semicircle)
Angle QPR = 18° (angle sum of triangle QPR)
$w = 18°$ (angles in same segment)

Exercise 12.4 (page 123)

1 $a = 90° - 40° = 50°$ (angle between tangent and radius)

2 $b = 50°$ (angle OXT = OYT = 90°, angle sum of quadrilateral)

3 $c = 70°$ (angle between tangent and radius)
$d = 20°$ (angles in same segment)

4 $6e = 90°$ (angle between tangent and radius)
$e = 15°$
$5e = 75°$
$f = 180° - 150° = 30°$ (isosceles triangle)

5 $g = 90° - 45° = 45°$ (angle between tangent and radius)

6 $h = 50°$ (angle sum of triangle)
$i = 30°$ (angle sum of triangle)

7 $j = 40°$ (angle between tangent and radius)
$k = 50°$ (angle sum of triangle)
$l = 40°$ (angles subtended by same arc)

8 Angle at O = 180° - 70° = 110°
Angle OXY = (180° - 110°) ÷ 2 = 35°
$m = n = 90° - 35° = 55°$ (angle between radius and tangent)

9 PY = 16 cm (perpendicular from centre to chord)
$a = 20$ cm (Pythagoras, 12, 16, 20 triangle)

10 OA = OB = 13 cm (radii)
Angle AMO = BMO = 90° (perpendicular from centre to chord)
OM = 25 - 13 = 12 cm
Therefore triangles AMO and BMO are congruent.
AM = 5 cm (Pythagoras, 5, 12, 13 triangle)
$b = $ AB = AM + MB = 5 + 5 = 10 cm

Graduated Assessment for OCR GCSE Mathematics © Hodder Murray 2007

Exercise 12.5 (page 126)

1. $a = 70°$, $b = 50°$ (angle between chord and tangent)

2. Triangle is equilateral
   $c = 60°$ and $d = 60°$ (angle between chord and tangent)

3. Angle CXA = 30°   (angles on straight line)
   Angle ABC = 150°   (opposite angles of cyclic quadrilateral)
   $e = f = 15°$ (base angles of isosceles triangle)

4. $g = 80°$ (angles at centre and circumference)
   $h = 180° - (90° + 80°) = 10°$ (angle between radius and tangent = 90°)
   $i = 180° - (90° + 40°) = 50°$ (angle at circumference = 40°)

5. $j = 80°$ (angle between chord and tangent)
   $k = 60°$ (angle sum of triangle)

6. $l = 60°$ (angle between chord and tangent)
   $m = 35°$ (angle between chord and tangent)

7. $n = 65°$ (angle sum of triangle)
   $o = 65°$ (angle between chord and tangent)

8. Tangents from same point are equal in length therefore $p = 60°$ (equilateral triangle)
   Therefore $q = 60°$ (angle between chord and tangent)
   Equilateral triangle inscribed in circle therefore $r = 60°$

9. $s = 70°$ (angle between chord and tangent)
   $t = 40°$ (isosceles triangle)

10. Angle ACB = u (angle between chord and tangent)
    Angle ABC = u (isosceles triangle)
    $4u = 180°$ (angle sum of triangle)
    $u = 45°$

11. Angle RAC = 61° (isosceles triangle)
    Angle QAB = 58° (isosceles triangle)
    $v = 61°$ (angles on a straight line)
    Angle BPC = 58° (angle sum of a triangle)
    Angle PBC = 61° (isosceles triangle)
    Angle QBA = angle QAB = 58°
    $w = 61°$ (angles on a straight line)
    $x = 58°$ (angle sum of a triangle)

12. Angle ABC = 85° (angles on a straight line)
    $y = 85°$ (angle between chord and tangent)
    $z = 71°$ (angle between chord and tangent)

13. TD = TA   (tangents from T)
    UA = UB   (tangents from U)
    VB = VC   (tangents from V)
    WC = WD   (tangents from W)
    Also TU = TA + AU = TD + UB
          VW = VC + CW = VB + WD
    Therefore TU + WV = (TD + UB) + (VB + WD)
               = (TD + DW) + (UB + VB)
               = WT + UV

13 Straight-line graphs

Exercise 13.1 (page 130)

1.
   a) $y = 3x + 2$
   b) $y = \bar{}x + 4$
   c) $y = 5x$
   d) $y = 4x - 1$
   e) $y = \bar{}2x + 5$
   f) $y = 3x$

2.
   a) $y = 4x + 2$
   b) $y = \frac{1}{3}x + 4$
   c) $y = 2x$
   d) $y = 2x + 3$
   e) $y = 4x$
   f) $y = \frac{1}{2}x + 1$

3.
   a) $y = \bar{}x + 5$
   b) $y = \bar{}1 \cdot 5x + 1$
   c) $y = \frac{\bar{}1}{2}x - 3$
   d) $y = \bar{}3x + 6$
   e) $y = \bar{}5x$
   f) $y = \bar{}2 \cdot 5x - 5$

4.
   a) $m = 4, c = \bar{}6$
   b) $m = \bar{}9, c = 7$
   c) $m = 1, c = 0$
   d) $m = \frac{1}{3}, c = \bar{}1\frac{1}{3}$
   e) $m = \bar{}0 \cdot 2, c = 1 \cdot 8$
   f) $m = \bar{}2, c = 5 \cdot 5$

5. $C = 84x$

6. a)

   b) 1·6 dollars/£; rate of currency conversion
   c) Flat rate charge of 5 dollars
   d) $d = 1 \cdot 6p - 5$

Exercise 13.2 (page 133)

1. Parallel line of form $y = 2x + c$
   Perpendicular line of form $y = \bar{}0 \cdot 5x + c$

2. Parallel line of form $y = \bar{}2x + c$
   Perpendicular line of form $y = 0 \cdot 5x + c$

3. $\frac{\bar{}1}{3}, 3$

4. $5, \frac{\bar{}1}{5}$

5. $y = 6x + 4$

6. $y = 3x + 2$

7. $y = \bar{}2x + 3$

8. $y = \bar{}4x - 1$

9 $\frac{1}{2}$

10 $\frac{^-1}{6}$

11 $x + 3y = 16$

12 $3y = 2x + 9$

13 a) $y = 4x + 3$ and $4x - y = 5$

    b) $2y - 3x = 5$ and $6y + 4x = 1$

14 $3y = x + 4$

15 a) $y = 2x + 2$

    b) $x + 2y = 4$

    c) $(0, 2)$

14 Surveys and sampling

Exercise 14.1 (page 137)

1 Check students' answers.

2 a) Possible answers: biased, only asking tennis players, and about tennis courts.

    b) Check students' answers.

Exercise 14.2 (page 139)

1 a) No – e.g. mainly male.

    b) No – likely to exclude families with small children.

    c) No – excludes people who work.

    d) No – would only include those who work in a particular area.

    Representative sample obtained by e.g. systematic sampling.

2 Check students' answers.

3 Check students' answers.

4 a) Possible answers: localised response, only asking people who are not at work. Better to visit different locations/areas and extend the hours.

    b) May not provide the right distribution of age and sex of students; better to choose by stratified or random sampling.

5 a) and b)

6 Check students' answers.

Exercise 14.3 (page 141)

1 Total number of students = 972
    Y7 = 21·4 i.e. 21   Y8 = 19·86 i.e. 20
    Y9 = 20·27 i.e. 20   Y10 = 19·5 i.e. 20
    Y11 = 18·9 i.e. 19

2 Total number of students = 464,
    Y11 = 25

3 (If P = population)
    Day 1 catch 16·7% implies $P = 2695$
    Day 2 catch 15·8% implies $P = 2848$
    Day 3 catch 20% implies $P = 2250$
    Day 4 catch 18·8% implies $P = 2394$
    Mean value of $P = 2547$

4 Check students' answers.

Exercise 14.4 (page 144)

1 a) Label the rows and columns 0 to 9, so that 0,0 represents a time of 14 seconds and 3,3 a time of 46 seconds.
    Use 10 random numbers to select the times of 10 students and calculate the mean and median for the sample and hence for the 100 students.

    b) Check students' answers; mean = 42·09; median = 35

2 a) Simple random sampling

    b) Stratified random sampling

3 a) Method 1
    The sample squares are 30, 63, 77, 52, 43, 00, 08, 03, 80 and 60.
    Mean number of worms = 53 ÷ 10 = 5·3
    Method 2
    The starting value is 04.
    The sample squares are 04, 14, 24, 34, 44, 54, 64, 74, 84 and 94.
    Mean number of worms = 43 ÷ 10 = 4·3

    b) They are quite close.

    c) Population mean = 487 ÷ 100 = 4·87

    d) The two sample means are quite close to the population mean; one is above and one below the population mean. The average of the two values is very close to the population mean value.

4 Year 9: 16
    Year 10: 16
    Year 11: 16
    Year 12: 6
    Year 13: 6

5 Total number of workers = 4010
    Factory 1: 10
    Factory 2: 5
    Factory 3: 50
    Factory 4: 25
    Factory 5: 10

Exercise 14.5 (page 147)

Possible answers

1 a) Only people who read that newspaper can reply. These may not be representative of the population.
    Only people who can be bothered to reply will have their views represented.

    b) Only finds the opinion of those who already use the car park. These may not be representative of the population. They may refuse to take part.

    c) The views of the eleventh house may be very different from those of the tenth and this will distort the results.

    d) Saturday morning is not a representative time of the week.

2 a) Excludes all those who are ex-directory, those without a phone and young people.

b) Only get people who are using the train at that time. These may not be representative of the population.

People may not want to take part as they are rushing home from work.

c) Likely to cover only people who read a lot.

Not a random selection.

Revision exercise D1 (page 149)

1

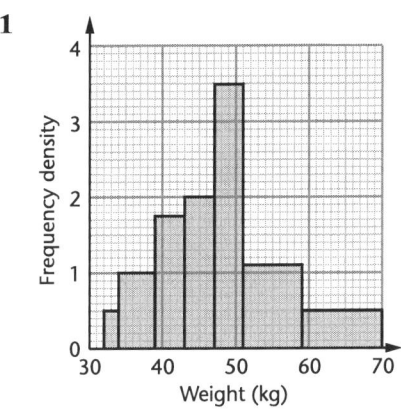

2 50

3

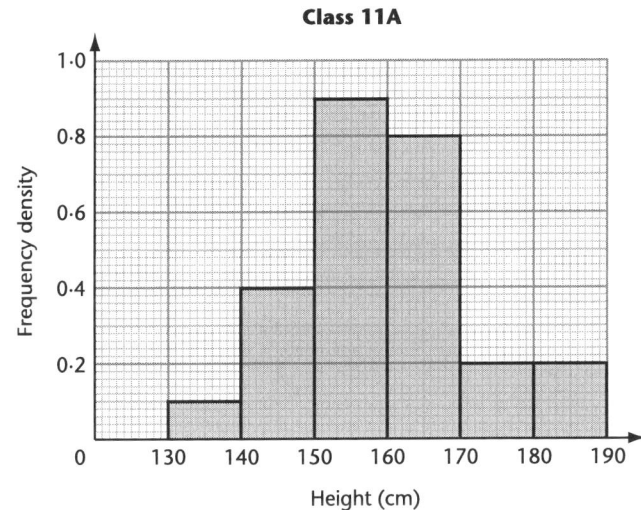

Class 11A

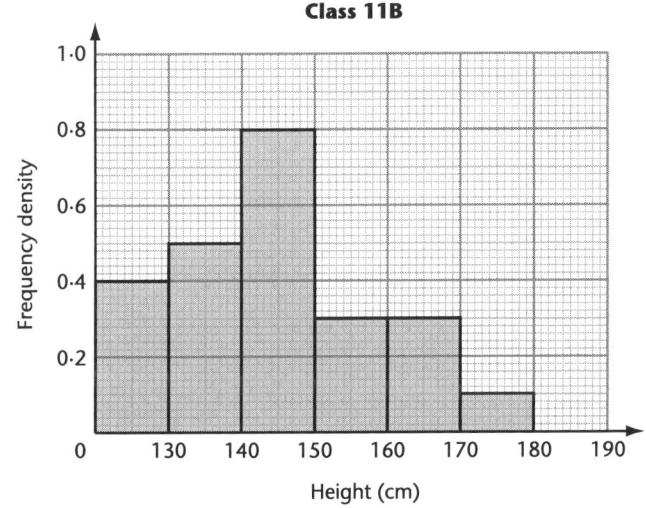

Class 11B

STAGE
9

Answers

4 a) $a = 49°$ (angle at the circumference $= \frac{1}{2}$ angle at the centre)

$b = 131°$ (sum of angles on a straight line)

b) $c = 120°$ (opposite angles of a cyclic quadrilateral)

$d = 240°$ (angles at centre and circumference)

$e = 120°$ (angles at centre and circumference)

c) Angle at centre $= 160°$ (angles at centre and circumference)

$f = 10°$ (base angle of isosceles triangle)

d) Angle DAB $= 70°$ (base angle of isosceles triangle)

Angle OAB = angle DAB – angle EAO
$= 70° – 40° = 30°$

Therefore angle AOB $= 120°$ (angle sum of isosceles triangle)

Therefore $g = 60°$ (angles at centre and circumference)

e) Angle at B $= 70°$ (base angle of isosceles triangle)

Therefore $h = 70°$ (angle between chord and tangent)

f) $i = 55°$ (base angle of isosceles triangle)

Therefore $j = 70°$ (angle sum of triangle)

$k = 55°$ (angle between chord and tangent)

$l = 70°$ (angle between chord and tangent)

g) Third angle of triangle $= 70°$ (angle between chord and tangent)

Therefore $m = 70°$ (base angle of isosceles triangle)

$n = 40°$ (angle sum of a triangle)

h) $o = 25°$ (angles in same segment)

Third angle of triangle $= 80°$ (vertically opposite angles)

$p = 75°$ (angle sum of a triangle)

i) $q = 62°$ (angles at centre and circumference)

$r = 28°$ (base angles of isosceles triangle)

$s = 29°$ (angle sum of a triangle)

j) $t = 5·66\,\text{cm}$ to 2 d.p. (Pythagoras)

$u = 38·9°$ to 1 d.p. (trigonometry)

5 a) $y = \frac{1}{2}x + 2$ or $2y = x + 4$

b) $y = \frac{-15}{8}x + 15$ or $15x + 8y = 120$

c) $y = \frac{5}{3}x - 1$ or $3y = 5x - 3$

6 a) $y = 3x + 2$

b) $y = {}^{-}2x + 6$ or $2x + y = 6$

c) $y = \frac{1}{2}x + 5$ or $2y = x + 10$

7 a) $y = \frac{5}{2}x - 10$ or $2y = 5x - 20$

b) $y = \frac{2}{5}x + \frac{8}{5}$ or $5y = 2x + 8$

c) $3x + y = 9$

8 AB: $x + 3y = 14$
BC: $y = 2x - 7$
AC: $5x + y = 14$

9 a) $y = 7x - 34$

b) $2y = 5x - 20$

c) $5x + 4y = 6$

10 a) $x + 2y = 4$

b) $3x + 2y = 2$

c) $2y = 5x - 11$

11 Yes – random sample of subscribers.

12 a) Random sample but limited to those with library card.

b) Stratified sample, 1% of each group.

13 a) No – because the number of pupils is significantly different.

b) 14

14 8, 12, 31, 6, 3

15 No – using whole crop from one tree, though this has been selected at random. A better sample would be one apple from each of 50 trees selected at random.

16 a) Number all the members.
Use a random number table or the random number generating facility on a calculator to select 50 of the numbers.

b) If there were 200 members, for example, then 1 in 4 need to be selected.
Use a random number table to choose a number between 1 and 4.
Starting with that number, keep adding 4. This will give the required 50 members for the sample.

c) Divide the members into groups, by age or sex, for example.
Work out the proportion of the 50 needed from each group.
Select these randomly using systematic or simple random sampling.

Stage 10 Contents

STAGE

10

Using graphs to solve equations

A10.5 (part)

Construct graphs of . . . the circle $x^2 + y^2 = r^2$; solve problems involving the intersection of straight lines with a curve (including a circle).

H2/6e, 6f, 6h

Objectives

- Solve simultaneous equations where one equation is quadratic
- Solve problems involving the intersection of a straight line with a curve, including a circle
- Recognise the equation of a circle, $x^2 + y^2 = r^2$

Prior knowledge

- Use calculators with trigonometrical and exponential functions
- Apply Pythagoras' theorem
- Plot the graphs of linear, quadratic, cubic and reciprocal functions
- Solve a pair of simultaneous linear equations graphically
- Solve a quadratic equation graphically

Equipment needed

2 mm graph paper, pairs of compasses

Speed-up sheets available

1.1, 1.2, 1.3

General notes

The remaining part of the module statement, involving exponential functions, is dealt with in the next chapter.

Students learned how to solve simultaneous linear equations graphically in Stage 8; this chapter develops the topic, looking first at examples where one of the two equations is a quadratic.

Tell students that, when solving equations graphically, they should give their answers to the nearest small division on the graph, usually 1 decimal place, as anything more accurate is impossible from a graph.

The next section develops the work done on solving quadratic equations in Stage 8. There the solution was found by finding the points of intersection of the curve with $y = 0$. Here students need to find the equation of a line of the form $y = c$ or $y = mx + c$ which they can then draw on the graph of the curve. The curve and line act as simultaneous equations and the solution is found by reading the x-coordinates of the points of intersection. Once students understand that they need to rearrange the equation of the curve they are given so that one side is the same as the equation of the curve they have drawn, or that has been drawn for them, and that the other side is the equation of the line they have to draw, they should not have a problem with this topic.

Questions on this topic usually include an easy part where the line is $y = 0$ or a whole number, and a harder part where a straight line needs to be found. In the examples and exercises there are some questions that test only the finding of these straight lines, to save drawing a large number of graphs, but the examination is likely to involve drawing the curve as well.

The final section of the chapter introduces the equation of a circle, using Pythagoras' theorem to explain its form. Make sure that students notice the tip explaining how they can draw the graph of a circle using compasses. At this level, all the circles students encounter will have their centre at the origin. Students may have to solve a problem involving the intersection of a circle and a line but they will be given the equation of the line to draw.

Graduated Assessment for OCR GCSE Mathematics © Hodder Murray 2007

Growth and decay

2

N10.1

Use calculators to explore exponential growth and decay.

H2/3t

A10.5 (part)

Construct graphs of exponential functions . . . ; solve problems involving the intersection of straight lines with a curve.

H2/6e, 6f

Objectives

- Draw graphs of exponential growth and decay
- Solve exponential equations graphically and using trial and improvement

Prior knowledge

- Deal with negative powers

Equipment needed

2 mm graph paper

Speed-up sheets available

2.1, 2.2

General notes

The chapter begins with a revision of the use of index notation in this context. It is essential here that students are able to use their calculators to evaluate any number to any power (positive or negative). Since they will sometimes be working with non-integral powers in this chapter, they cannot use repeated multiplication in many of the examples they will meet. If students are not confident in this, it may be worthwhile giving them an exercise in it. It is fairly easy to make up an exercise of questions such as $2^{3.2}$, 5^{-3}, $2 \cdot 5^3$ although, at this level, it is unlikely to be necessary for most students.

Once students have mastered this, the exercises should be relatively straightforward. The difficult part is spotting the exponential rules but, with practice, the fact that a constant multiplier leads to an equation of the form

start value $\times$ (multiplier)^x

should become fairly routine.

Students need to be able to recognise the graphs of exponential functions, and the next section of the chapter deals with this. Students also learn that they can solve exponential equations graphically using the same method as they would to solve quadratic equations graphically. Students should, however, be able to see that it is more difficult to get a good estimate of the solution to the problem when graphs of exponential equations are involved: this leads into solving exponential equations using trial and improvement.

Students who ask whether there isn't a better method than trial and improvement could be pointed towards logarithms as another way of dealing with indices. However, using a multiplier and the powers key on a calculator is the required method in the National Curriculum for exploring growth and decay.

STAGE
10

Growth and decay

As extension work, you may like to use this problem.

James puts £200 into his savings account each year, at the beginning of the financial year.

He receives 5% compound interest each year.

How much will he have in his account at the end of the tenth year?

A spreadsheet could be used here.

	A	B	C	D	E	F
				D5 ▼ f_x =(D4 + 200)*1.05		
1	£200 invested each year at 5% compound interest					
2						
3		TIME		TOTAL		
4	end of year	1		£210.00		
5		2		£430.50	£200 added at start of year, so (210 + 200) × 1.05	
6		3		£662.03		
7		4		£905.13		
8		5		£1,160.38		
9		6		£1,428.40		
10		7		£1,709.82		
11		8		£2,005.31		
12		9		£2,315.58		
13		10		£2,641.36		
14						

Notes on the tasks

Activity 1 (page 168)

5%	15 years
1%	70 years
2%	36 years
10%	8 years

Graduated Assessment for OCR GCSE Mathematics © Hodder Murray 2007

Rational and irrational numbers

N10.2

Convert a recurring decimal to a fraction and vice versa; use prime factors to identify fractions which represent terminating decimals; simplify expressions involving powers or surds including rationalising a denominator.

H2/3c, 3n

Objectives

- Recognise rational and irrational numbers
- Use prime factors to identify fractions which represent terminating decimals
- Convert a recurring decimal to a fraction and vice versa
- Simplify expressions involving surds
- Simplify expressions by rationalising a denominator

Prior knowledge

- Recognise natural numbers and integers
- Recognise terminating decimals and recurring decimals
- Understand and use the 'dot' notation for recurring decimals
- Multiply out and simplify brackets such as $(a + b)(c + d)$

Equipment needed

None

General notes

The chapter starts by explaining what rational and irrational numbers are. Students need to be able to recognise rational numbers and irrational numbers. Effectively this simply means being able to recognise surds and functions of π; once students have grasped this, they should not find the exercise questions too demanding. The most common mistake in examinations is to identify recurring decimals as irrational.

You can explain how the different types of real number relate to each other using the standard notation and a Venn diagram.

$\mathbb{N}$ Natural numbers
$\mathbb{Z}$ Integers
$\mathbb{Q}$ Rational numbers
$\mathbb{R}$ Real numbers

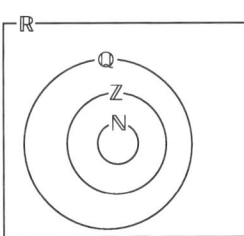

The next section looks at terminating and recurring decimals in more detail. In particular, students need to be able to identify which fractions will give recurring decimals and which will give terminating decimals by examining the prime factors of the denominator of the fraction. The denominator of a fraction giving a terminating decimal will have only 2 and/or 5 as its prime factors.

STAGE

10

The method of converting fractions into decimals by division is straightforward enough. The 'dot' notation, however, may need to be revised. Converting recurring decimals to fractions is a little harder. Students should learn results such as $0.\dot{2} = 0.222... = \frac{2}{9}$ and may also wish to learn results such as $0.\dot{3}\dot{4} = 0.343434... = \frac{34}{99}$ but results such as $0.2\dot{3}1\dot{4} = 0.231\,431\,431\,4...$ are more difficult and may need to be calculated using the method demonstrated in the Student's Book.

The final section in the chapter looks at surds and how to manipulate them. Some students find this work quite difficult and it does need practice. The exercises are necessarily fairly repetitive. It helps if the students are well versed in algebra in general and multiplication of brackets in particular. Students may find it helpful to learn these results, if they haven't already.

$$(a + b)^2 = a^2 + 2ab + b^2$$
$$(a - b)^2 = a^2 - 2ab + b^2$$
$$(a + b)(a - b) = a^2 - b^2$$

Notes on the tasks

Challenge 1 (page 182)

Assume $\sqrt{2}$ is rational.

Then $\sqrt{2}$ is $\frac{a}{b}$ for some integers a and b whose only common factor is 1.

$2 = \dfrac{a^2}{b^2}$ Squaring both sides.

$2b^2 = a^2$ (1)   Multiplying both sides by b^2.

Since a^2 has 2 as a factor, a^2 is even.
So a is also even.

Let $a = 2c$, where c is an integer.

$a^2 = 4c^2$ Squaring both sides.
$2b^2 = 4c^2$ Substituting for a^2 in equation (1).
$b^2 = 2c^2$

Since b^2 has 2 as a factor, b^2 is even.
So b is also even.

But if a and b are both even then a and b have a common factor, 2, and the original premise is false.

So $\sqrt{2}$ is not rational; it is irrational.

The proof that $\sqrt{2}$ is irrational may be of interest to very able students. It is based on the 'reductio ad absurdum' principle. You can introduce the ⇒ symbol to save students having to write 'implies that' or similar.

Activity 1 (page 183)
Sevenths

$\frac{1}{7} = 0.\dot{1}4285\dot{7}$ $\frac{2}{7} = 0.\dot{2}8571\dot{4}$ $\frac{3}{7} = 0.\dot{4}2857\dot{1}$

$\frac{4}{7} = 0.\dot{5}7142\dot{8}$ $\frac{5}{7} = 0.\dot{7}1428\dot{5}$ $\frac{6}{7} = 0.\dot{8}5714\dot{2}$

The six-digit pattern in each case includes the same six digits, in the same order cyclically.

You can tell at which point in the cycle the decimal starts for each fraction by arranging the digits in the cycle in numerical order: 1, 2, 4, 5, 7, 8. This is because each fraction is larger than the previous fraction.

Ninths

$\frac{1}{9} = 0.\dot{1}$ $\frac{2}{9} = 0.\dot{2}$ $\frac{3}{9} = 0.\dot{3}$

$\frac{4}{9} = 0.\dot{4}$ $\frac{5}{9} = 0.\dot{5}$ $\frac{6}{9} = 0.\dot{6}$

$\frac{7}{9} = 0.\dot{7}$ $\frac{8}{9} = 0.\dot{8}$

The pattern for ninths is obvious: the digit that recurs is the same as the numerator of the fraction. This could lead to a discussion about whether $0.\dot{9}$ is equal to 1.

Elevenths

$\frac{1}{11} = 0.\dot{0}\dot{9}$ $\frac{2}{11} = 0.\dot{1}\dot{8}$ $\frac{3}{11} = 0.\dot{2}\dot{7}$

$\frac{4}{11} = 0.\dot{3}\dot{6}$ $\frac{5}{11} = 0.\dot{4}\dot{5}$ $\frac{6}{11} = 0.\dot{5}\dot{4}$

$\frac{7}{11} = 0.\dot{6}\dot{3}$ $\frac{8}{11} = 0.\dot{7}\dot{2}$ $\frac{9}{11} = 0.\dot{8}\dot{1}$

$\frac{10}{11} = 0.\dot{9}\dot{0}$

Again, the pattern should be obvious. In each case, the two digits that recur are equal to the numerator of the fraction multiplied by 9.

Thirteenths

$\frac{1}{13} = 0.\dot{0}7692\dot{3}$ $\frac{2}{13} = 0.\dot{1}5384\dot{6}$ $\frac{3}{13} = 0.\dot{2}30769$

$\frac{4}{13} = 0.\dot{3}0769\dot{2}$ $\frac{5}{13} = 0.\dot{3}84615$ $\frac{6}{13} = 0.\dot{4}6153\dot{8}$

$\frac{7}{13} = 0.\dot{5}3846\dot{1}$ $\frac{8}{13} = 0.\dot{6}15384$ $\frac{9}{13} = 0.\dot{6}9230\dot{7}$

$\frac{10}{13} = 0.\dot{7}6923\dot{0}$ $\frac{11}{13} = 0.\dot{8}4615\dot{3}$ $\frac{12}{13} = 0.\dot{9}2307\dot{6}$

There are two six-digit cyclic patterns here. Pattern 1 consists of the digits 076 923; pattern 2 consists of the digits 153 846.

You can tell at which point in the cycle of which pattern the decimal starts for each fraction by arranging the digits in the cycle in numerical order. There are two 3s and two 6s: in each case, looking at the next digit in each pattern will determine which will equate to the larger fraction.

Fraction	$\frac{1}{13}$	$\frac{2}{13}$	$\frac{3}{13}$	$\frac{4}{13}$	$\frac{5}{13}$	$\frac{6}{13}$	$\frac{7}{13}$	$\frac{8}{13}$	$\frac{9}{13}$	$\frac{10}{13}$	$\frac{11}{13}$	$\frac{12}{13}$
Pattern 1	0		2	3					6	7		9
Pattern 2		1			3	4	5	6			8	

Seventeenths

$\frac{1}{17} = 0.\dot{0}5882352941176\dot{4}7$

$\frac{2}{17} = 0.\dot{1}1764705882352\dot{9}4$

$\frac{3}{17} = 0.\dot{1}7647058823529\dot{4}1$

$\frac{4}{17} = 0.\dot{2}3529411764705\dot{8}8$

$\frac{5}{17} = 0.\dot{2}9411764705882\dot{3}5$

$\frac{6}{17} = 0.\dot{3}5294117647058\dot{8}2$

$\frac{7}{17} = 0.\dot{4}1176470588235\dot{2}9$

$\frac{8}{17} = 0.\dot{4}7058823529411\dot{7}6$

$\frac{9}{17} = 0.\dot{5}2941176470588\dot{2}3$

$\frac{10}{17} = 0.\dot{5}8823529411764\dot{7}0$

$\frac{11}{17} = 0.\dot{6}4705882352941\dot{1}7$

$\frac{12}{17} = 0.\dot{7}0588235294117\dot{6}4$

$\frac{13}{17} = 0.\dot{7}6470588235294\dot{1}1$

$\frac{14}{17} = 0.\dot{8}2352941176470\dot{5}8$

$\frac{15}{17} = 0.\dot{8}8235294117647\dot{0}5$

$\frac{16}{17} = 0.\dot{9}4117647058823\dot{5}2$

There is one 16-digit cyclic pattern here. As before, you can tell at which point in the cycle the decimal starts for each fraction by arranging the digits in the cycle in numerical order, and looking at the next digit if there is more than one. Although most scientific calculators will not give 16 figures, the pattern can be used to fill in the gaps.

This task is a good opportunity for group work: students can work out a few fractions each and pool their results.

4 Trigonometry in non-right-angled triangles

S10.3

Calculate the area of a triangle using $\frac{1}{2}ab\sin C$; use the sine and cosine rules to solve 2-D and 3-D problems.

H3/2g

Objectives

- Determine when to use the sine rule or the cosine rule
- Calculate the area of a triangle using $\frac{1}{2}ab\sin C$

Prior knowledge

- Use Pythagoras' theorem and trigonometry in right-angled triangles

Equipment needed

None

General notes

Notation will be important in this chapter. Go through this carefully since the formulae derived depend on this notation.

Advise students that for an isosceles or an equilateral triangle it is not appropriate to use the sine or cosine rules. The triangle should be split in half and right-angled triangle trigonometry should be employed.

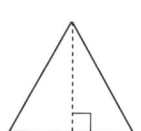 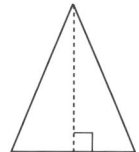

The formulae for the sine rule, where three fractions are equated, can be confusing. Emphasise the need to establish which form

to use (depending on whether a length or an angle is to be found) and demonstrate how the two relevant fractions are identified: they should contain the value to be calculated and three other known values.

Diagrams showing one example of each of the situations in which the sine rule can be used are included in the Student's Book. You may wish to ask students to draw the other possibilities (there are only two situations where the sine rule is used, but the orientation of the triangle may be different from that shown). Explain what is meant by the 'included' and the 'non-included' angle.

The formula for the cosine rule is given on the examination formulae sheet only in the form $a^2 = b^2 + c^2 - 2bc\cos A$. This can be rearranged to give $\cos A = \dfrac{b^2 + c^2 - a^2}{2bc}$, in order to find an angle, but students may prefer to substitute the values and rearrange it afterwards instead.

Try to make students aware of the pattern in the formulae, as this will help them remember the different forms. In particular, note the connection between the angle letter and the length with the same letter.

Mistakes are often made in questions involving the cosine rule because students do not evaluate the expression in the correct order, particularly the '$- 2bc\cos A$' part. Spend some time explaining the procedure. Brackets are used in the worked examples to emphasise the appropriate order of evaluation. Other problems arise when the angle substituted into the formula is obtuse. This leads to a double negative in the third term and can cause errors. Efficient use of a calculator is important and it is worth taking some time over this with your students.

The work on the area of a triangle is more straightforward than that on the sine and cosine rules. Again point out the pattern in the letters used (abC, bcA, caB) and how they form a circular structure. Make sure students understand that the formula requires two adjacent sides and the included angle.

Notes on the tasks

The activities in this chapter give opportunities for using ideas of proof. Depending on the group, just part **a)** of each activity can be used as a lead-in, giving practice with multi-step questions and showing that use of the cosine or sine rule or the formula for the area of a triangle turns unstructured work into a one-step problem. With more able groups or individual students, part **b)** of the activities can be used to practise algebra and show how part of the general result is obtained. Deriving the formulae for the sine rule, the cosine rule and the area of a triangle, however, is not a requirement of the course and these formulae will be given on the examination formulae sheet.

Activity 1 (page 193)

a) In triangle BCP $\quad \sin 70° = \dfrac{h}{5 \cdot 2}$

so that $\qquad\qquad h = 5 \cdot 2 \sin 70°$

In triangle ABP $\quad \sin 38° = \dfrac{h}{c}$

so that $\qquad\qquad \sin 38° = \dfrac{5 \cdot 2 \sin 70°}{c}$

$\qquad\qquad\qquad c = \dfrac{5 \cdot 2 \sin 70°}{\sin 38°}$

$\qquad\qquad\qquad c = 7 \cdot 9 \text{ cm (to 1 d.p.)}$

b) In triangle ABC the line BP is drawn so that angle APB is a right angle.

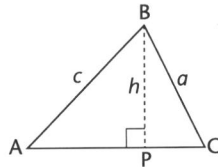

(i) In triangle BCP $\quad \sin C = \dfrac{h}{a}$

so that $\qquad\qquad h = a \sin C$

(ii) In triangle ABP $\quad \sin A = \dfrac{h}{c}$

so that $\qquad\qquad h = c \sin A$

(iii) Equating the two expressions for h gives

$a \sin C = c \sin A$

which can be rearranged to give

$$\frac{a}{\sin A} = \frac{c}{\sin C}$$

If you repeat this process with a line, CQ, drawn in triangle ABC so that angle CQA is a right angle, then you can establish the relationship

$$\frac{a}{\sin A} = \frac{b}{\sin B}$$

Combining these gives the sine rule.

$$\frac{a}{\sin A} = \frac{b}{\sin B} = \frac{c}{\sin C}$$

As is stated in the Student's Book, this formula can be inverted to give an alternative form.

$$\frac{\sin A}{a} = \frac{\sin B}{b} = \frac{\sin C}{c}$$

Activity 2 (page 198)

a) In triangle BCP, using Pythagoras' theorem gives

$$a^2 = h^2 + (8 \cdot 2 - x)^2$$

so that $\qquad a^2 = h^2 + 67 \cdot 24 - 16 \cdot 4x + x^2 \quad (1)$

In triangle ABP $\quad h^2 + x^2 = 5 \cdot 1^2$

Also in triangle ABP $\quad \cos 50° = \dfrac{x}{5 \cdot 1}$

so that $\qquad\qquad x = 5.1 \cos 50°$

Substituting for $(h^2 + x^2)$ and x in equation (1) gives

$$a^2 = 5 \cdot 1^2 + 67 \cdot 24 - 16 \cdot 4 \times 5 \cdot 1 \cos 50°$$

$$a = \sqrt{5 \cdot 1^2 + 67 \cdot 24 - 83 \cdot 64 \cos 50°}$$

$$a = 6 \cdot 3 \text{ cm (to 1 d.p.)}$$

STAGE
10

b) In triangle ABC, the line BP is drawn so that angle BPC is a right angle.

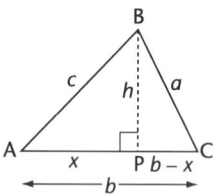

(i) In triangle BCP, using Pythagoras' theorem gives

$$a^2 = h^2 + (b-x)^2$$

so that $\quad a^2 = h^2 + b^2 - 2bx + x^2 \quad (1)$

(ii) In triangle ABP $\;c^2 = h^2 + x^2$

(iii) Also in triangle ABP $\quad \cos A = \dfrac{x}{c}$

so that $\quad x = c \cos A$

(iv) Substituting for $(h^2 + x^2)$ and x in equation (1) gives

$$a^2 = c^2 + b^2 - 2bc \cos A$$

which is usually written as

$$a^2 = b^2 + c^2 - 2bc \cos A$$

Similarly, if a line AQ or CR is drawn then the equivalent formulae can be derived.

$$b^2 = a^2 + c^2 - 2ac \cos B$$

and $\;c^2 = a^2 + b^2 - 2ab \cos C$

These formulae can be rearranged into alternative forms that may be used when finding angles.

$$\cos A = \frac{b^2 + c^2 - a^2}{2bc}$$

$$\cos B = \frac{c^2 + a^2 - b^2}{2ca}$$

$$\cos C = \frac{a^2 + b^2 - c^2}{2ab}$$

Activity 3 (page 204)

a) In triangle BCP $\;\sin 63° = \dfrac{h}{5\cdot8}$

so that $\quad\quad\quad h = 5\cdot8 \sin 63°$

Since $\quad$ Area of a triangle $= \frac{1}{2} \times$ base $\times$ height

then Area of triangle ABC $= \frac{1}{2} \times 7\cdot6 \times h$

$\quad\quad\quad\quad\quad\quad\quad = \frac{1}{2} \times 7\cdot6 \times 5\cdot8 \sin 63°$

$\quad\quad\quad\quad\quad\quad\quad = 19\cdot6 \,\text{cm}^2$ (to 1 d.p.)

b) In triangle ABC, the line BP is drawn so that angle APB is a right angle.

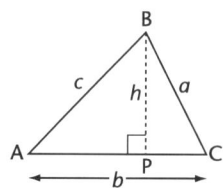

(i) You know that

$\quad\quad\quad$ Area of triangle $= \frac{1}{2} \times$ base $\times$ height

$\quad\quad\quad\quad\quad\quad\quad\quad = \frac{1}{2} bh \quad\quad (1)$

In triangle BCP $\;\sin C = \dfrac{h}{a}$

so that $\quad\quad\quad h = a \sin C$

(ii) Substituting for h in equation (1) gives

$\quad\quad$ Area of triangle $= \frac{1}{2} ba \sin C$

This is more easily remembered as 'half $ab \sin C$'.

So $\quad$ Area of triangle $= \frac{1}{2} ab \sin C$

In a similar way we can show that there are two other equivalent formulae.

$\quad\quad$ Area of a triangle $= \frac{1}{2} bc \sin A$

and $\;$ Area of a triangle $= \frac{1}{2} ca \sin B$

Graduated Assessment for OCR GCSE Mathematics © Hodder Murray 2007

Trends and time series

D10.2

Identify seasonality and trends in time series, from tables or diagrams; interpret graphs modelling real situations.

H2/6d, H4/5b

Objectives

- Identify seasonality and trends in time series
- Interpret graphs that model real situations

Prior knowledge

- Plot time series graphs
- Calculate moving averages
- Interpret simple graphs of everyday situations, such as a journey

Equipment needed

Access to the internet, 2 mm graph paper

Speed-up sheets available

5.1

General notes

The interpretation of time series graphs was touched on in Stage 8, for completeness, but properly belongs here. This chapter gives further practice on moving averages, with the emphasis this time on interpreting the graphs rather than producing them. It also extends the work to making predictions based on the data.

ICT may be used to produce further examples, but there are some problems. Excel, for instance, plots the moving average points for its trendline at the end of the data period being used, instead of at the centre. However, this problem may be circumvented by specifying the coordinates for the moving averages as a separate series.

The module statement also includes interpretation of graphs modelling real situations. The chapter includes examples of distance–time and velocity–time graphs as well as questions on water filling a container or bath. For an additional activity on this topic, use graphs students have seen or generated in their study of other subjects.

Notes on the tasks

Activity 1 (page 212)

The graphs students find could be used for class discussion. They may also find data that could be used to produce additional examples.

6 Congruency – proving and using

S10.2

Understand and use SSS, SAS, ASA and RHS condition to prove the congruence of triangles; verify standard ruler and compass constructions; use congruence to show that translations, reflections and rotations preserve length and angle.

H3/1e, 2e, 3b

Objectives

- Prove that triangles are congruent and understand and use the notation SSS, SAS, ASA and RHS when proving congruency
- Verify the standard ruler and compass methods of construction
- Use congruency to show that translation, rotation and reflection preserve length and angle size

Prior knowledge

- Draw the perpendicular bisector of a line
- Draw the bisector of an angle

Equipment needed

Pairs of compasses, protractors or angle measurers

General notes

Students should know the basic ideas of congruency from much earlier in the course. The opening activity leads towards the different ways in which a triangle can be defined uniquely and hence to the formal conditions for congruency.

These conditions are then used in showing why formal constructions such as bisecting a line produce the result they do. There are also applications to the properties of quadrilaterals and to transformations, such as the position of the centre of a rotation.

Notes on the tasks

Activity 1 (page 218)

a) All the triangles will be congruent.
b) This is the ambiguous case.
Two triangles are possible: one with an acute angle at C and one with an obtuse angle.
c) (i) The triangle in part **a)** is an example of the SSS condition.
The other three conditions which can lead to one triangle only are SAS, ASA and RHS, as the Student's Book goes on to say.
(ii) The triangle in part **b)** is one example of the ambiguous case.
Two triangles can result for a triangle given any two sides and either of the non-included angles.

Activity 2 (page 224)

a), b) Check students' diagrams.

c) Two triangles are formed by drawing the radii from A and B to the points where the construction arcs cross. These points are labelled C and D.

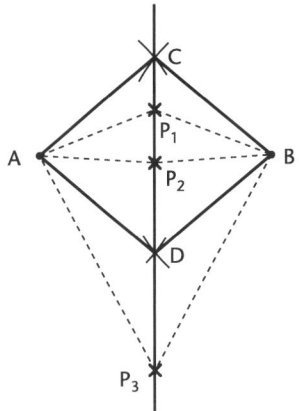

AC = BC (radii), AD = BD (radii) and CD is common.
So triangles ACD and BDC are congruent (SSS).
So angle ACD = angle BCD.

At any point (P) on the line CD:
AC = BC, angle ACP = angle BCP and CP is common.
So triangles ACP and BCP are congruent (SAS).
So AP = PB.

This proves that at any point on the perpendicular bisector of the line joining two points, A and B, the distance from that point to A is equal to the distance from that point to B.

As well as practising logical argument, this activity is also useful revision of the construction of the perpendicular bisector of a line.

Activity 3 (page 224)

a) Check students' diagrams.

b) Two triangles are formed by drawing the radii from the points where the first pair of construction arcs cross the lines AB and AC (labelled D and E) to the point where the second pair of construction arcs cross (labelled F).

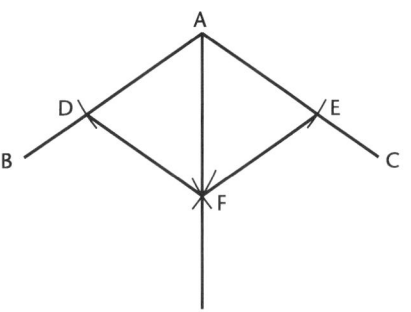

AD = AE (radii), DF = EF (radii) and AF is common.
So triangles ADF and AEF are congruent (SSS).
So angle DAF = angle EAF.

Since the radius chosen for the arcs was arbitrary, F could be at any point on the angle bisector.

Again, this practises logical argument and revises the construction of the bisector of an angle.

STAGE
10

7 Calculating the roots of equations

A10.2

Solve quadratic equations by completing the square and using the quadratic formula.

H2/5k

Objectives

- Solve quadratic equations by completing the square
- Solve quadratic equations using the formula

Prior knowledge

- Expand and simplify $(ax + b)(cx + d)$
- Factorise quadratic expressions
- Solve quadratic equations using factorising or graphs

Equipment needed

None

General notes

Before tackling the first section of this chapter, students need to be able to manipulate algebraic expressions with confidence and to recognise the patterns which arise in a 'complete square' quadratic expression. Activity 1 will help them to brush up on their expansion skills and to find the patterns if they have not noticed them before, but more practice may be needed.

Example 2 shows an equation to which a fraction has to be added to complete the square. The alternative method given for this example uses a multiplier to avoid having to deal with fractions; this method is

also used in Example 3. It would also be possible to solve Example 3 by dividing through by 2 to make the coefficient of the x^2 term 1: you may prefer to teach this method.

As stated in the Student's Book, errors are often made with the signs when using the formula, especially with ^-4ac. Some students may find it easier to work out $b^2 - 4ac$ before substituting the numbers into the formula.

Another common error is failure to divide the whole expression by $2a$: students sometimes divide just part of the expression by $2a$ or use just 2 even when $a \neq 1$.

To avoid these errors, encourage students to write down their working: they should begin by stating the values of a, b and c. When they evaluate the expression, they need either to use brackets or to press the equals key before dividing by $2a$ (which, if they are entering as it $2 \times a$, must itself be in brackets). It is probably safer to work out and write down the value of $2a$.

Notes on the tasks

Activity 1 (page 227)

$$(x + 1)^2 = (x + 1)(x + 1) = x^2 + 2x + 1$$
$$(x - 2)^2 = (x - 2)(x - 2) = x^2 - 4x + 4$$
$$(x + 3)^2 = (x + 3)(x + 3) = x^2 + 6x + 9$$
$$(x - 5)^2 = (x - 5)(x - 5) = x^2 - 10x + 25$$
$$(2x + 1)^2 = (2x + 1)(2x + 1) = 4x^2 + 4x + 1$$
$$(3x - 2)^2 = (3x - 2)(3x - 2) = 9x^2 - 12x + 4$$
$$(5x + 4)^2 = (5x + 4)(5x + 4) = 25x^2 + 40x + 16$$

$$(x - 4)^2 = x^2 - 8x + 16$$
$$(4x + 1)^2 = 16x^2 + 8x + 1$$
$$(2x - 3)^2 = 4x^2 - 12x + 9$$

Students should notice the following when expanding the expression $(mx + k)^2$.

- The coefficient of x^2 is m^2.
- The coefficient of x is twice the product of m and k.
- The constant term is k^2.

Activity 2 (page 227)

$x^2 + 4x + 4 = (x + 2)^2$
$x^2 + 6x + 9 = (x + 3)^2$
$x^2 + 10x + 25 = (x + 5)^2$
$x^2 + 2x + 1 = (x + 1)^2$
$x^2 + 12x + 36 = (x + 6)^2$
$x^2 - 6x + 9 = (x - 3)^2$
$x^2 - 16x + 64 = (x - 8)^2$
$x^2 - 7x + 12 \cdot 25 = (x - 3 \cdot 5)^2$
$x^2 + 5x + 6 \cdot 25 = (x + 2 \cdot 5)^2$
$4x^2 + 12x + 9 = (2x + 3)^2$
$9x^2 + 6x + 1 = (3x + 1)^2$
$6x^2 + 24x + 24 = 6(x + 2)^2$

To complete the factorisations students should use the patterns discovered in Activity 1.

Activity 3 (page 230)

a) 0, 3   **b)** 5   **c)** ⁻2
d) 4, ⁻2   **e)** ⁻6, 1   **f)** $n, -m$

This task illustrates one of the other main uses of completing the square: finding the minimum or maximum value of a quadratic function.

Challenge 1 (page 231)

The hint given in the Student's Book suggests multiplying through to avoid fractions.

$ax^2 + bx + c = 0$
$4a^2x^2 + 4abx + 4ac = 0$ — Multiply through by $4a$.
$4a^2x^2 + 4abx + 4ac + b^2 = b^2$ — Add b^2 to both sides.
$4a^2x^2 + 4abx + b^2 = b^2 - 4ac$ — Subtract $4ac$ from both sides.
$(2ax + b)^2 = b^2 - 4ac$ — The left-hand side is now a perfect square.
$2ax + b = \pm\sqrt{b^2 - 4ac}$ — Take the square root of both sides.
$2ax = -b \pm \sqrt{b^2 - 4ac}$ — Subtract b from both sides.
$x = \dfrac{-b \pm \sqrt{b^2 - 4ac}}{2a}$ — Divide both sides by $2a$.

The alternative method is to divide through so the coefficient of x^2 is 1.

$ax^2 + bx + c = 0$
$x^2 + \dfrac{bx}{a} + \dfrac{c}{a} = 0$ — Divide both sides by a.
$x^2 + \dfrac{bx}{a} + \dfrac{c}{a} + \dfrac{b^2}{4a^2} = \dfrac{b^2}{4a^2}$ — Add $\dfrac{b^2}{4a^2}$ to both sides.
$x^2 + \dfrac{bx}{a} + \dfrac{b^2}{4a^2} = \dfrac{b^2}{4a^2} - \dfrac{c}{a}$ — Subtract $\dfrac{c}{a}$ from both sides.
$\left(x + \dfrac{b}{2a}\right)^2 = \dfrac{b^2}{4a^2} - \dfrac{c}{a}$ — The left-hand side is now a perfect square.
$\left(x + \dfrac{b}{2a}\right)^2 = \dfrac{b^2 - 4ac}{4a^2}$ — Put the right-hand side over $4a^2$.
$x + \dfrac{b}{2a} = \dfrac{\pm\sqrt{b^2 - 4ac}}{2a}$ — Take the square root of both sides.
$x = \dfrac{-b \pm \sqrt{b^2 - 4ac}}{2a}$ — Subtract $\dfrac{b}{2a}$ from both sides.

Calculating the roots of equations

STAGE 10

Graduated Assessment for OCR GCSE Mathematics © Hodder Murray 2007   73

Surface areas and complex shapes

S10.1

Solve problems involving surface areas and volumes of pyramids, cylinders, cones and spheres, and problems involving more complex shapes including segments of circles and frustums of cones.

H3/2i

Objectives

- Solve problems involving surface areas and volumes of pyramids, cylinders, cones and spheres
- Solve problems involving complex shapes, including segments of circles and frustums of cones

Prior knowledge

- Find the circumference and area of a circle
- Find the arc length and area of a sector
- Find the area of a triangle
- Find the volume of a prism, pyramid, cone or sphere
- Rearrange formulae
- Use Pythagoras' theorem and trigonometry

Equipment needed

Models of three-dimensional shapes (optional), A3 paper, pairs of compasses, scissors, sticky tape, protractors or angle measurers

General notes

This chapter provides opportunities to revise and apply topics studied earlier. Revisiting the formulae is particularly useful as so few of these are given on the formulae sheet in examinations. There are also opportunities to extend the work beyond the strict confines of the specification. For example, in Challenge 1 students are invited to prove for themselves the formula for the curved surface area of a cone.

As with all three-dimensional work, models of pyramids, cones and other solids can help students who find visualisation difficult in interpreting the context. Some of the more complex problems may seem daunting at first, but such questions can be broken down into smaller stages and will become less challenging with practice.

Notes on the tasks

Activity 1 (page 234)

a)

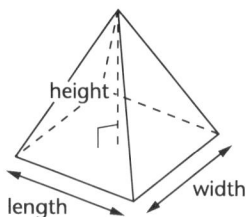

The dimensions required in order to calculate the volume are the length and width of the base and the perpendicular height, as marked on the diagram.

b)

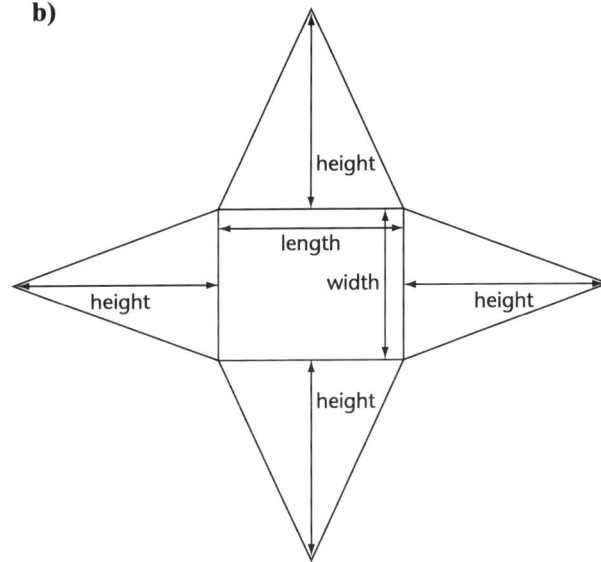

The dimensions required in order to calculate the surface area are the length and width of the base and the perpendicular heights of the triangular faces, as marked on the diagram.

The perpendicular heights of the triangular faces are not the same as the perpendicular height of the pyramid but can be worked out from this measurement and the dimensions of the base.

Activity 2 (page 235)

b) The arc length of the sector, and therefore the circumference of the base of the cone, will be $\frac{\theta}{360} \times 2 \times \pi \times 12 = \frac{\theta\pi}{15}$, where θ is the angle chosen by the student.

d) The radius of the base of the cone will be $\frac{\text{Answer to part b)}}{2\pi}$.

e) The area of the sector, and therefore the surface area of the cone, will be $\frac{\theta}{360} \times \pi \times 144 = \frac{20\theta\pi}{5}$.

This task involves actually making a cone and, as such, will help students understand how the base radius is formed, as well as the relationship between the dimensions of a sector and the cone made from it.

Challenge 1 (page 235)

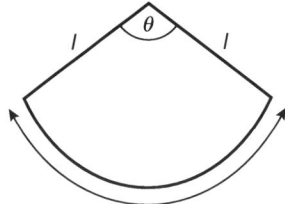 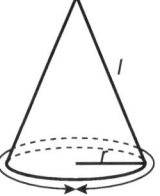

Arc length of sector $= \frac{\theta}{360} \times 2 \times \pi \times l$

This becomes the circumference of the base of the cone made from this sector.

Circumference of the base of the cone $= 2 \times \pi \times r$

$$\frac{\theta}{360} \times 2 \times \pi \times l = 2 \times \pi \times r$$

$$\frac{\theta}{360} \times l = r$$

So the radius, r, of the base of a cone of slant height l can be expressed as $r = \frac{\theta}{360} \times l$.

Area of sector $= \frac{\theta}{360} \times \pi \times l^2$

This becomes the curved surface area of the cone made from this sector.

Since, for a cone, $r = \frac{\theta}{360} \times l$, the curved surface area can be expressed as $A = \pi r l$.

This task extends the work done in Activity 2 to look at the general results for cones.

Challenge 2 (page 244)

Curved surface area of cylinder $= 2\pi r h$

Surface area of cube $= 6r^2$

So $2\pi r h = 6r^2$

$$h = \frac{6r^2}{2\pi r}$$

$$= \frac{3r}{\pi}$$

Surface areas and complex shapes

STAGE
10

9 Working with algebraic fractions

A10.1

Manipulate algebraic expressions including fractions; solve related equations.

H2/5a, 5b, 5f

Objectives

- Manipulate algebraic expressions, including those with fractions
- Solve equations, including those involving algebraic fractions

Prior knowledge

- Add, subtract, multiply, divide and simplify numerical fractions
- Simplify and factorise algebraic expressions
- Rearrange formulae

Equipment needed

None

General notes

Once the basic concept of writing algebraic fractions with a common denominator is understood, most mistakes are due to skipping a step or missing a negative sign. It is worth stressing that only factors of the full numerator and denominator can be cancelled, and that factors can be numbers, letters or brackets containing both letters and numbers.

Solving equations involving algebraic fractions and brackets requires a good grasp of how to manipulate algebraic expressions. Sometimes students fail to multiply both sides of an equation by the common denominator, especially when one side is just a whole number.

Notes on the tasks

Activity 1 (page 249)

a) (i) $1\frac{7}{24}$ (ii) $1\frac{1}{10}$

(iii) $\frac{119}{120}$ (iv) $\frac{23x}{24}$

(v) $\frac{13x}{45}$

b) (i) $\frac{1}{4}$ (ii) $\frac{2}{3}$

(iii) $\frac{x}{3}$ (iv) $\frac{2}{9x}$

(v) $\frac{3}{2a}$

Vectors

S10.5

Understand and use vector notation; calculate, and represent graphically the sum of two vectors, the difference of two vectors and a scalar multiple of a vector; calculate the resultant of two vectors; understand and use the commutative and associative properties of vector addition; solve simple geometrical problems in 2-D using vector methods.

H3/3f

Objectives

- Understand and use vector notation
- Find the scalar multiple of a vector
- Calculate and represent the sum and difference of two vectors
- Calculate and represent the resultant of two vectors
- Solve simple geometrical problems in two dimensions using vector methods

Prior knowledge

- Describe translations using vector notation
- Add, subtract, multiply, divide and simplify numerical fractions
- Simplify and factorise algebraic expressions
- Rearrange formulae

Equipment needed

Squared paper

Speed-up sheets available

10.1

General notes

Students learned how to describe a translation using a vector earlier in the course and should find the work on column vectors reasonably straightforward. It is important to remind them that a column vector must be a column without a fraction line, not a pair of coordinates. They will be penalised in examinations if they do not write column vectors correctly.

Many students find vector geometry difficult. With general vectors, there are three important points that should be emphasised.

- If two lines are parallel then their vectors will be multiples of each other.
- If two lines are parallel and equal in length then their vectors will be equal.
- The vector $\overrightarrow{AB}$ in a geometrical shape is equal to any route from A to B around the shape.

Some students miss out some of the working when finding vectors and this often leads to errors.

This is a topic on which students may encounter multi-step questions in the examinations. With such questions it is especially important to show working as it can be worth a number of marks, all of which will be lost if the answer is incorrect and no working is shown.

STAGE

10

Notes on the tasks

Activity 1 (page 256)

a)

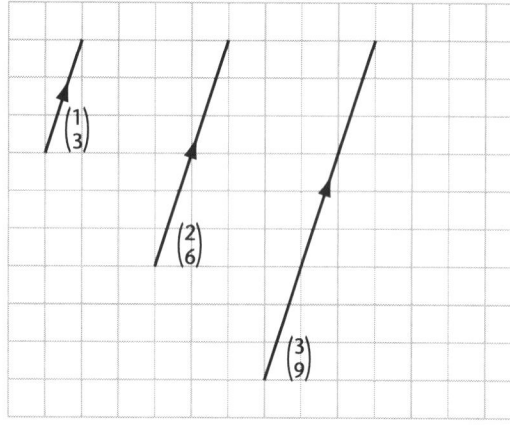

b)

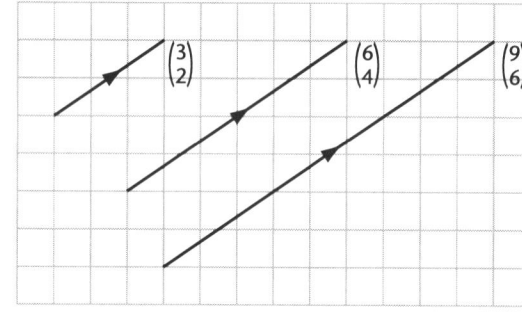

In both cases, the three vectors are parallel.

Activity 2 (page 267)

Check students' answers.

The sum of the vectors used for each route will always be $\begin{pmatrix} 7 \\ 0 \end{pmatrix}$.

There are a large number of possible routes here, especially if you allow a point to be visited more than once.

Less able students may need help finding their first route and understanding that the column vectors for the route add up to the column vector for $\overrightarrow{AE}$.

STAGE
10

Vectors

Comparing sets of data

D10.1

Compare data sets (including grouped discrete and continuous data); draw conclusions.

H4/5d

Objectives

- Compare data sets and draw conclusions from data, relating comparisons to the context of the data

Prior knowledge

- Plot and interpret cumulative frequency curves and box plots
- Plot and interpret histograms
- Find the mean, median and mode
- Find the range, quartiles and interquartile range

Equipment needed

Newspapers and magazines, 2 mm graph paper

Speed-up sheets available

11.1

General notes

This chapter draws together work done earlier in the course, with the emphasis on comparing data.

Interpreting graphs drawn by others is an important skill and so activities and a question on newspaper graphs are included.

As well as the tools in the National Curriculum, you may also have taught your students standard deviation. This will not be examined and is not included here but is, of course, a useful tool in comparing data. Informal ideas of skewness may also be useful but again are not required by the specification. There is a short introduction to skewness in this chapter, which you may omit if you wish.

The exercises provide practice in drawing cumulative frequency graphs, box plots and histograms, to refresh these skills, as well as asking for interpretation.

Notes on the tasks

Activity 1 (page 275)

This activity emphasises the need to plan what data is required in order to answer a question, and how to collect it.

Activity 2 (page 277)

a) (i) The greatest 3-month average of domestic car sales in the US was approximately 10·7 million, and occurred in 1973–4.
   (ii) 12 million sales
   (iii) Possible answers include: the scales are small and difficult to read accurately; having two vertical scales on one graph is confusing.
b) Check students' questions and answers.

Answering the questions composed by the pairs or groups could be done as a class activity.

Activity 3 (page 278)

Discussion of these graphs could be done in a plenary session.

STAGE
10

12 Simultaneous equations

A10.3

Solve exactly, by elimination of an unknown, two simultaneous equations in two unknowns, one of which is linear, the other equation quadratic in one unknown or of the form $x^2 + y^2 = r^2$.

H2/5I

Objectives

- Solve two simultaneous equations in two unknowns, one of which is linear and the other quadratic
- Solve two simultaneous equations in two unknowns, one of which is linear and the other of the form $x^2 + y^2 = r^2$

Prior knowledge

- Solve linear simultaneous equations algebraically
- Solve quadratic equations algebraically
- Solve quadratic equations graphically
- Recognise the equation of a circle radius r, with centre the origin, as $x^2 + y^2 = r^2$

Equipment needed

None

General notes

In Stage 8, students used the method of elimination to solve simultaneous equations algebraically. Many will prefer this method for solving simultaneous linear equations, and this is perfectly acceptable. Where one equation is a quadratic or the equation of a circle, and therefore involves powers of x and/or y, however, the method of substitution must be used. Previously students have dealt with such cases graphically, but they also need to be able to solve them algebraically.

Do not allow students to be confused by the module statement: 'elimination of an unknown' means writing one of the unknowns in terms of the other.

The module statement speaks of exact solutions and most of the resulting quadratic equations will factorise. However, the opportunity has been taken, in questions 13 to 18 of Exercise 12.3, to include questions where the resulting quadratic equation does not factorise. This provides an opportunity for revision of the methods of completing the square and using the quadratic formula. Students have been asked to give their answers to 2 decimal places, a degree of accuracy greater than would be sensible with graphical solution of such equations. Strict interpretation of the module statement would require students to leave their answers in surd form. As surds were covered in Chapter 3, you may also wish to practise this.

Notes on the tasks

Activity 1 (page 289)

a) $x = 4, y = 1$
b) $x = 3, y = 2$
c) $x = 1, y = 2$
d) $x = 3, y = -2$
e) $x = \frac{1}{2}, y = 1$

Trigonometrical functions

S10.4

Draw, sketch and describe the graphs of trigonometric functions for angles of any size, including transformations involving scalings in either or both the x and y directions.

H3/2g

Objectives

- Describe, draw and sketch the graphs of the sine, cosine and tangent functions for angles between 0° and 360°
- Describe, draw and sketch the graphs of other trigonometrical functions for angles between 0° and 360°

Prior knowledge

- Use a calculator with trigonometrical functions
- Use Pythagoras' theorem

General notes

The activities provide opportunities for an investigational approach to the topic, whilst the text uses a formal approach to defining cosine, sine and tangent for all angles. An alternative is for students to use the trigonometrical functions on their calculator to derive values of, for example, $\sin x$, to make a table of values and hence to draw the graph of $y = \sin x$ for $0° \leqslant x \leqslant 540°$. It is perhaps worth doing one such graph manually so that students can appreciate how it is generated, which they may not do so readily if it just appears on the screen using ICT.

The emphasis in this module statement is on the graphs of the functions. This has been extended here to using the symmetries of the graphs to find values of x that satisfy an equation such as $\sin x = 0\cdot5$. This helps students to focus on the symmetries, but may be omitted for weaker students, although some discussion of the symmetries of the graphs is important, as is the use of the terms *period* and *amplitude*, as these will enable students to describe the graphs as required.

Equipment needed

Graphics calculator, graph-drawing software or 2 mm graph paper

Speed-up sheets available

13.1, 13.2

STAGE

10

Notes on the tasks

Activity 1 (page 300)

$\sin^{-}40° = ^{-}0{\cdot}643$ $\sin^{-1}{}^{-}0{\cdot}643 = ^{-}40°$
$\sin 120° = 0{\cdot}866$ $\sin^{-1}0{\cdot}866 = 60°$
$\sin 270° = ^{-}1$ $\sin^{-1}{}^{-}1 = ^{-}90°$
$\sin 300° = ^{-}0{\cdot}866$ $\sin^{-1}{}^{-}0{\cdot}866 = ^{-}60°$

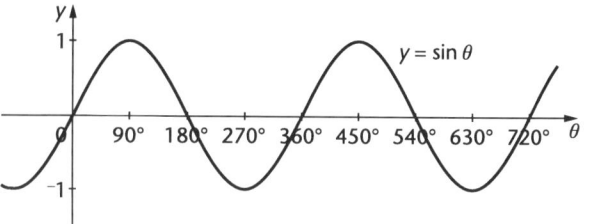

Students should reach the following conclusions.

$90° < \theta < 180°$	Only $\sin\theta$ is positive, $\cos\theta$ and $\tan\theta$ are negative.
$180° < \theta < 270°$	Only $\tan\theta$ is positive.
$270° < \theta < 360°$	Only $\cos\theta$ is positive.
$^{-}90° < \theta < 0°$	Only $\cos\theta$ is positive.
$^{-}180° < \theta < ^{-}90°$	Only $\tan\theta$ is positive.
$^{-}180° < \theta < ^{-}270°$	Only $\sin\theta$ is positive.
$^{-}270° < \theta < ^{-}360°$	All positive.

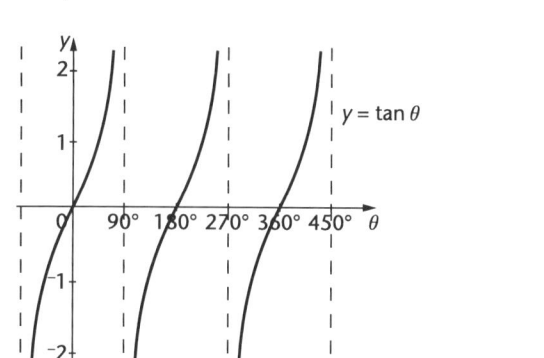

A calculator always gives the value for the inverse of sine, cosine and tangent in these ranges.

Cosine: $0° \leqslant \theta \leqslant 180°$
Sine: $^{-}90° \leqslant \theta \leqslant 90°$
Tangent: $^{-}90° < \theta \leqslant 90°$

Activity 2 (page 305)

a)

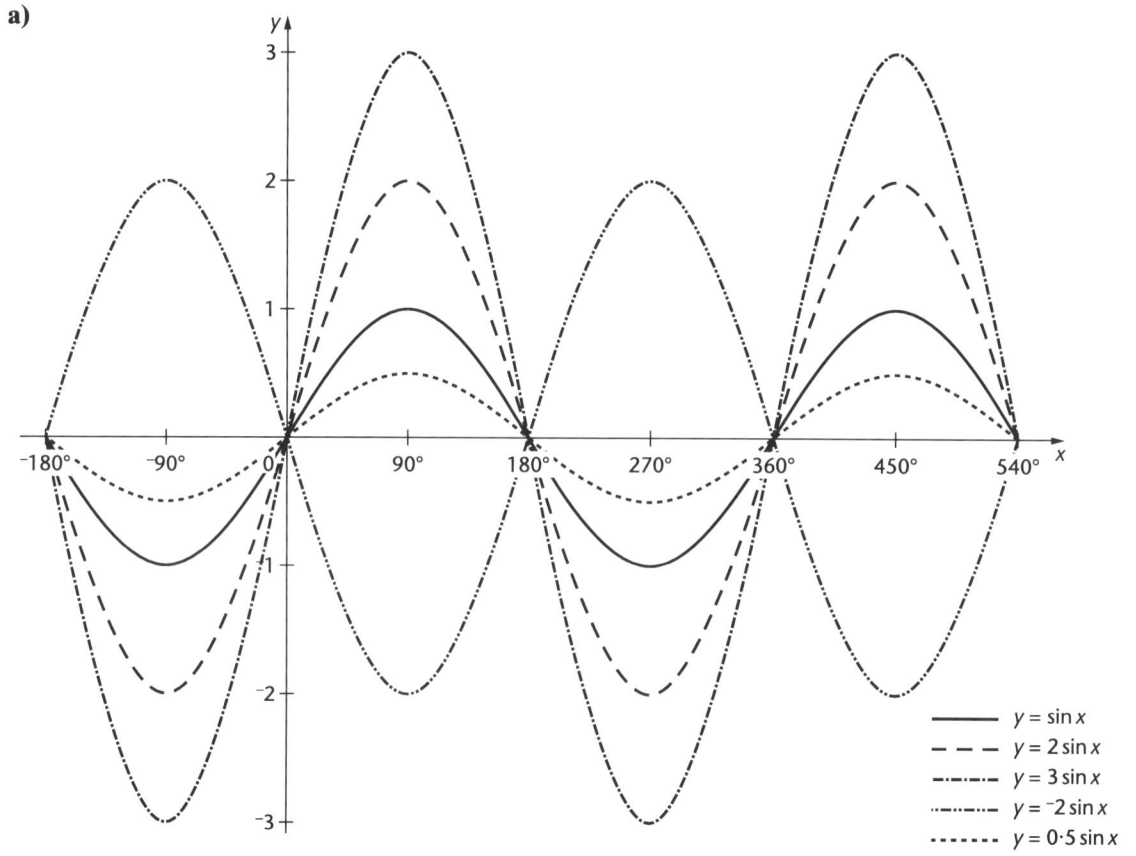

The graph of $y = a\sin x$ has the same period as the graph of $y = \sin x$ but the amplitude is multiplied by a.

Graduated Assessment for OCR GCSE Mathematics © Hodder Murray 2007

b)

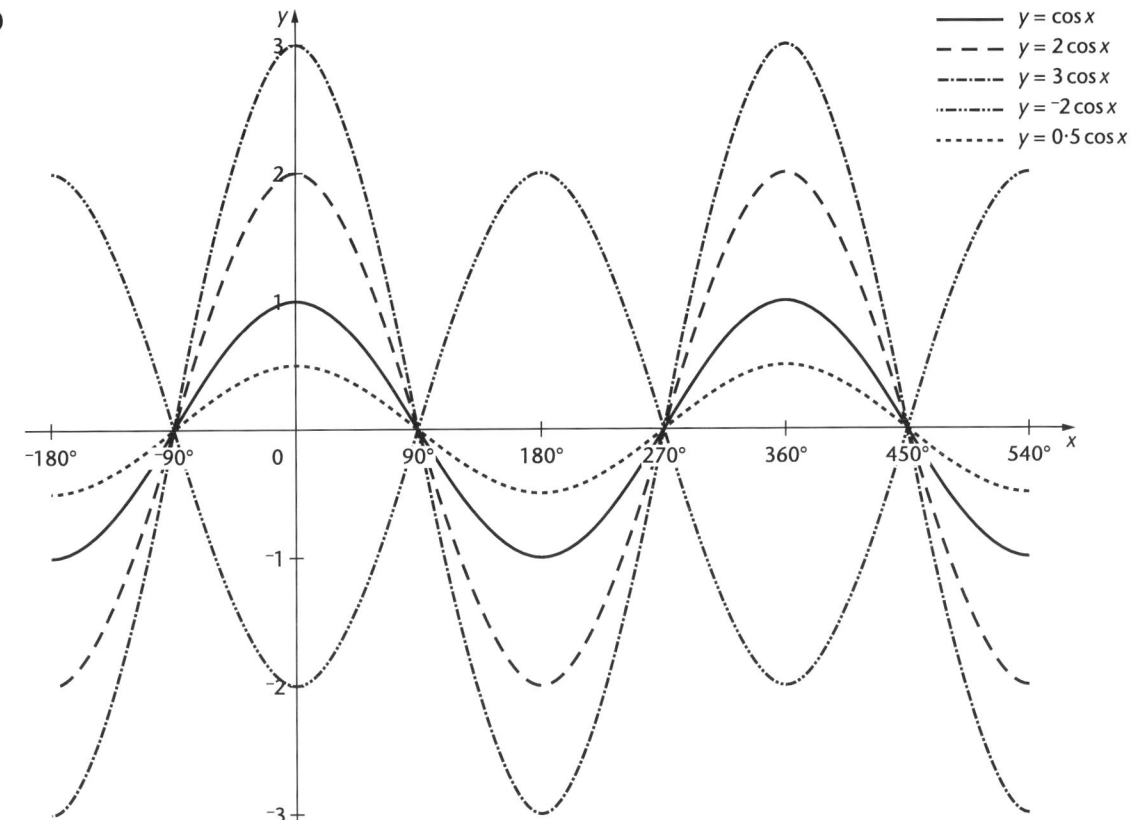

The graph of $y = a\cos x$ has the same period as the graph of $y = \cos x$ but the amplitude is multiplied by a.

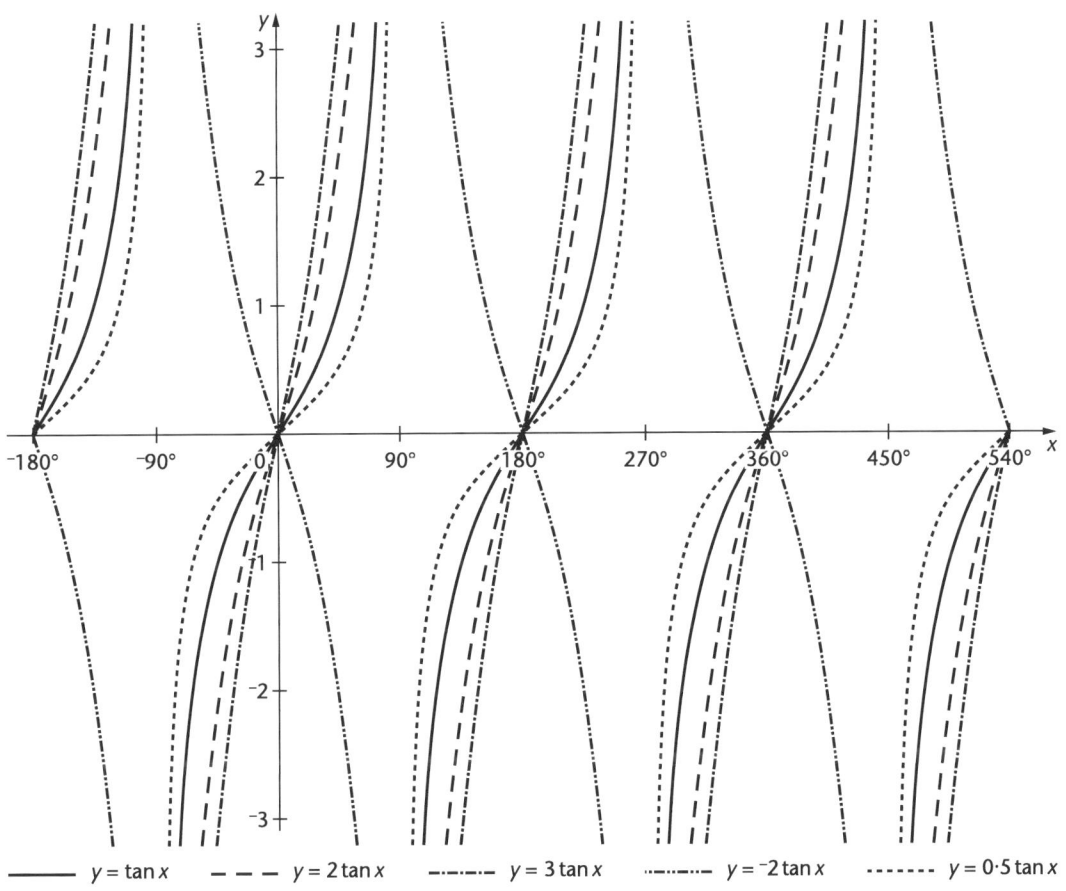

In a similar way, the graph of $y = a\tan x$ has asymptotes in the same places but the values of y are multiplied by a.

c)

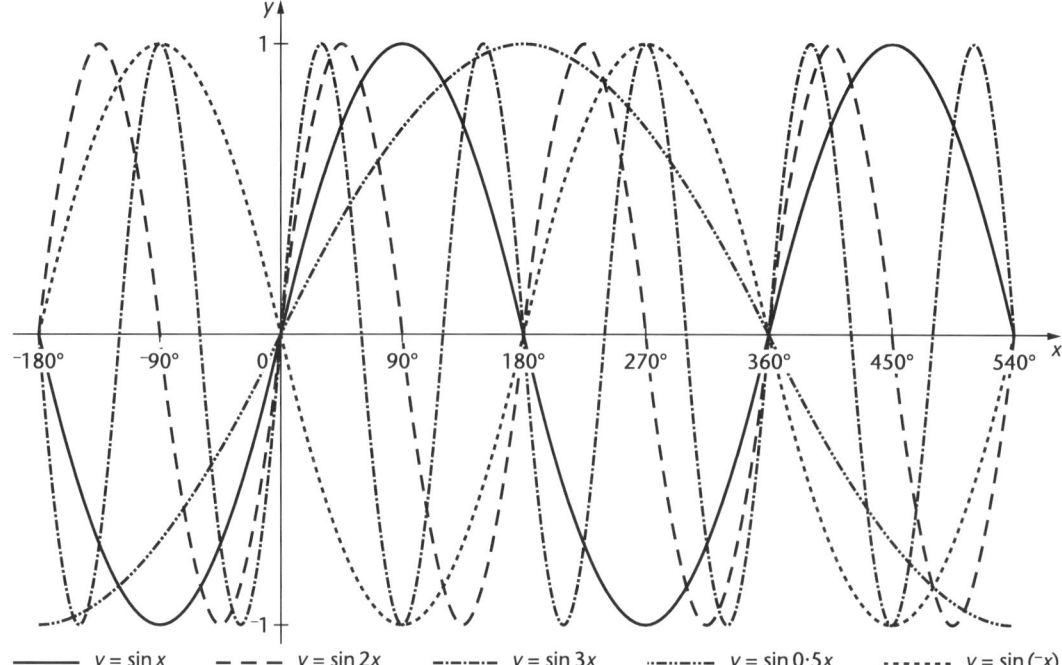

—— $y = \sin x$	– – – $y = \sin 2x$	–·–·– $y = \sin 3x$	···—··· $y = \sin 0{\cdot}5x$	······ $y = \sin(^-x)$

The graph of $y = \sin bx$ has the same amplitude as the graph of $y = \sin x$ but the period is $\dfrac{360°}{b}$.

d)

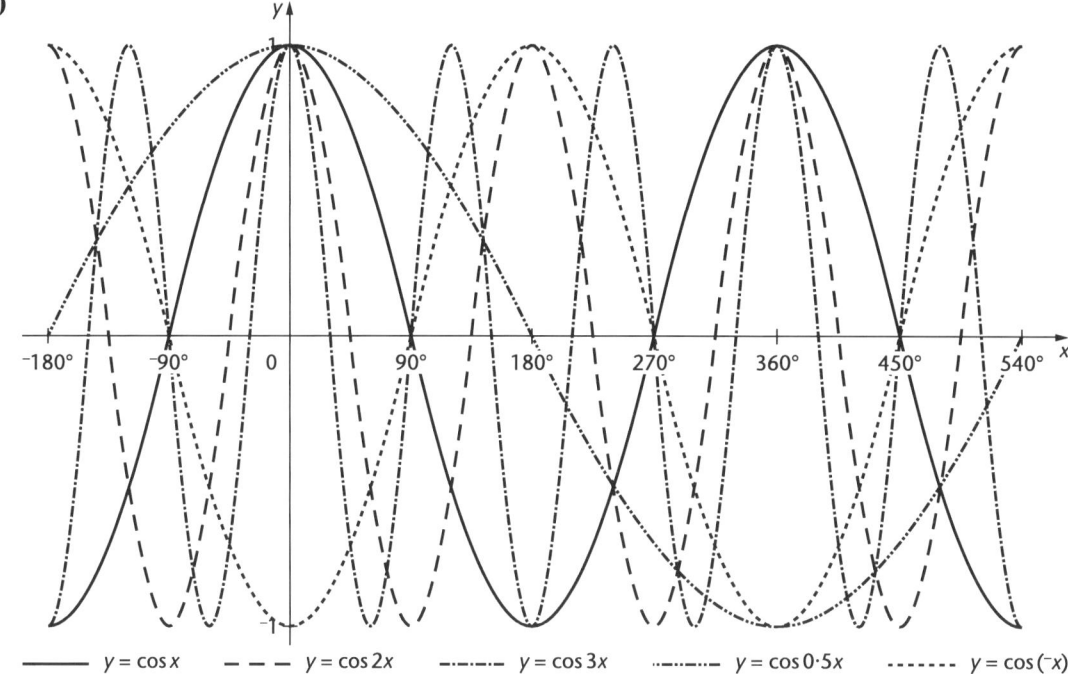

—— $y = \cos x$	– – – $y = \cos 2x$	–·–·– $y = \cos 3x$	···—··· $y = \cos 0{\cdot}5x$	······ $y = \cos(^-x)$

The graph of $y = \cos bx$ has the same amplitude as the graph of $y = \cos x$ but the period is $\dfrac{360°}{b}$.

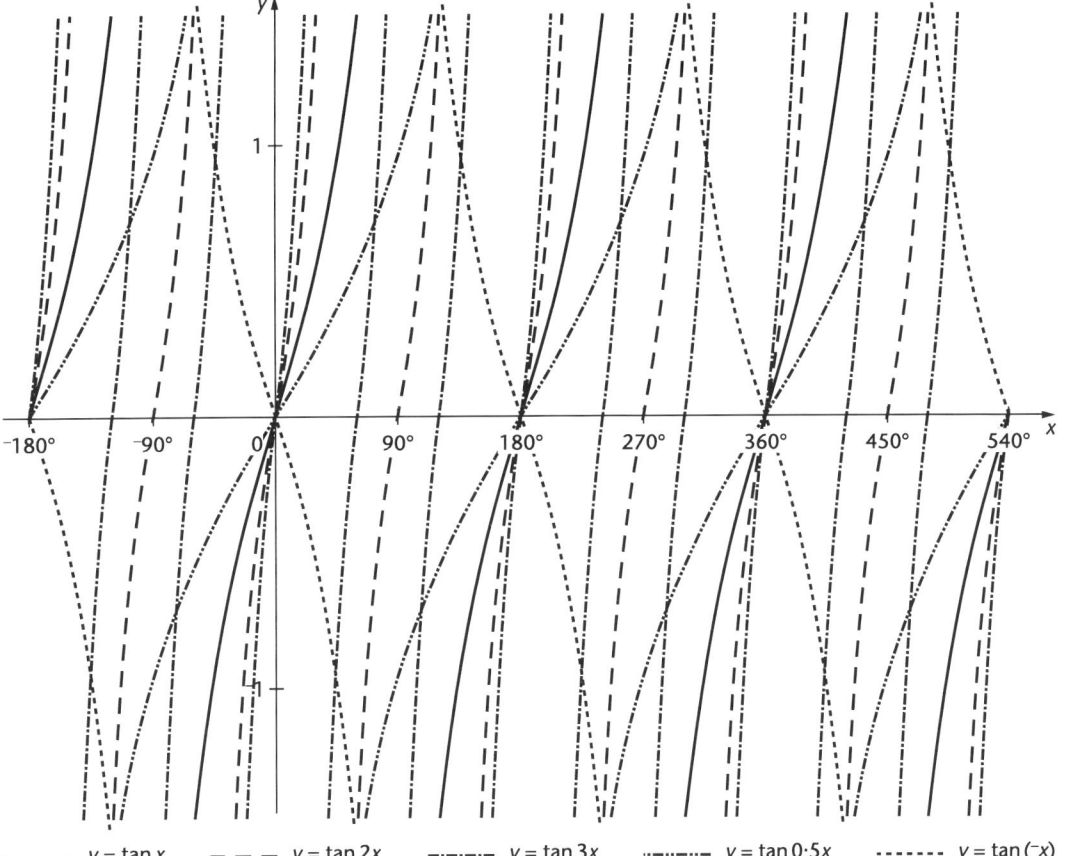

——— $y = \tan x$ – – – $y = \tan 2x$ –·–·–· $y = \tan 3x$ ···–···–· $y = \tan 0.5x$ ······ $y = \tan (^-x)$

In a similar way, the graph of $y = \tan bx$ has the same y values as the graph of $y = \tan x$, but the asymptotes are closer together (for $b > 1$) or further apart (for $b < 1$).

This task uses ICT to introduce the work on transformations, helping students to compare graphs for themselves.

14 Transforming functions

A10.4

Apply to the graph of $y = f(x)$ the transformations $y = f(x) + a$, $y = f(ax)$, $y = f(x + a)$, $y = af(x)$, for linear, quadratic, sine and cosine functions $f(x)$.

H2/6g

Objectives

- Carry out the transformations $y = f(x) + a$, $y = f(ax)$, $y = f(x + a)$ and $y = af(x)$ on linear, quadratic, sine and cosine graphs

Prior knowledge

- Carry out and describe reflections and translations
- Find the equation of a straight line
- Recognise the shapes of graphs such as $y = x^2$, $y = x^3$ and $y = \sin x$

Equipment needed

Graph-drawing software or 2 mm graph paper

General notes

This chapter develops the work on transforming graphs which was begun in Chapter 13.

Function notation is introduced briefly, to enable generalising. The Programme of Study specifies linear, quadratic, sine and cosine graphs, but the transformations also apply to other graphs and students could also explore them: cubic or reciprocal graphs would be a good place to start.

Notes on the tasks

Both the activities use ICT. A graph-drawing program such as Omnigraph is probably a better option than graphics calculators, since printouts may easily be obtained. If appropriate technology is not available and the graphs must be produced by hand, the activities may be adapted so that students can work in groups.

Activity 1 (page 310)

Section 1

a)

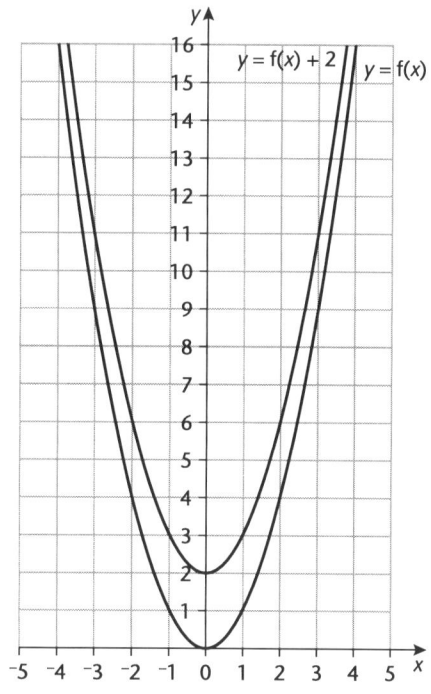

b) Translation of $\begin{pmatrix} 0 \\ 2 \end{pmatrix}$.

c) Check students' graphs.

d) Translation of $\begin{pmatrix} 0 \\ a \end{pmatrix}$.

Graduated Assessment for OCR GCSE Mathematics © Hodder Murray 2007

Section 2

a)

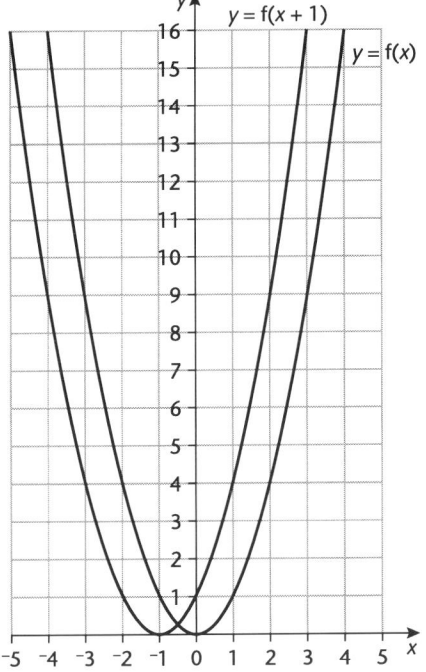

y = f(x + 1)

y = f(x)

b) Translation of $\begin{pmatrix} -1 \\ 0 \end{pmatrix}$.

c)

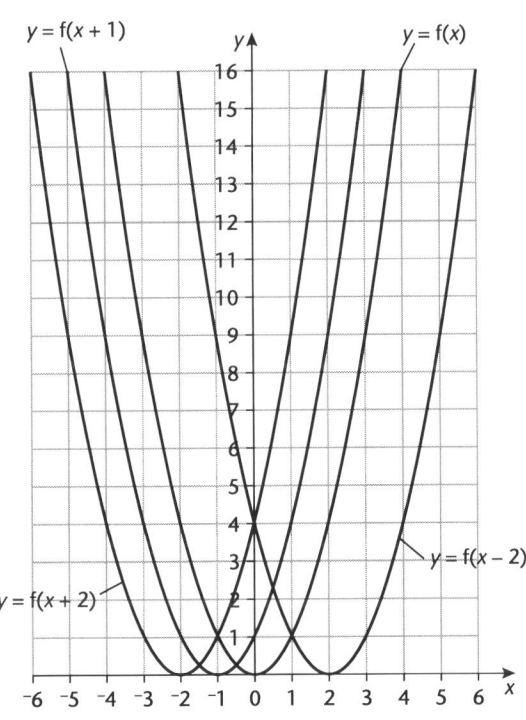

y = f(x + 1)

y = f(x)

y = f(x − 2)

y = f(x + 2)

d) Check students' graphs.

Translation of $\begin{pmatrix} -a \\ 0 \end{pmatrix}$.

Section 3

The same is true for any function.

The graph of f(x) + a is the graph of f(x) translated by $\begin{pmatrix} 0 \\ a \end{pmatrix}$.

The graph of y = f(x + a) is the graph of y = f(x) translated by $\begin{pmatrix} -a \\ 0 \end{pmatrix}$.

Activity 2 (page 315)

Section 1

a)

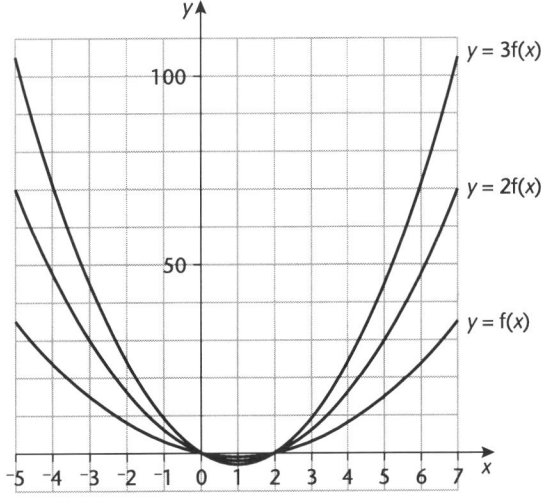

y = 3f(x)

y = 2f(x)

y = f(x)

b) Check students' graphs.

The transformation that maps the graph of y = f(x) on to y = kf(x), for any value of k, is a one-way stretch of scale factor k, parallel to the y-axis.

STAGE
10

Section 2

a)

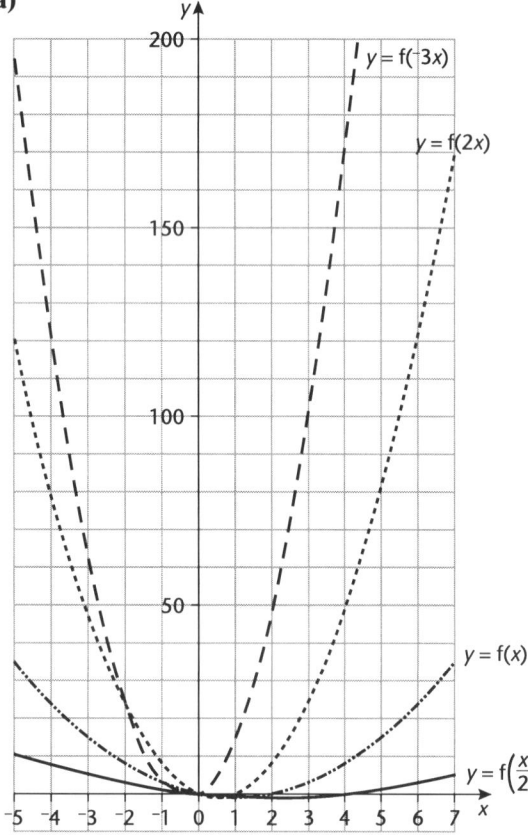

b) Check students' graphs.

The transformation that maps the graph of $y = f(x)$ on to $f(kx)$, for any value of k, is a one-way stretch of scale factor $\frac{1}{k}$, parallel to the x-axis.

Students may find part **b)** quite difficult. As well as drawing the graphs of $f(kx)$ for further values of k, they could compare the x values for various values of y. A good place to start is the lowest point on the curve.

For $f(x) = x^2 - 2x$, the lowest point occurs when $y = {}^-1$.

For $f(x)$, when $y = {}^-1$, $x = 1$.
For $f(2x)$, when $y = {}^-1$, $x = \frac{1}{2}$.
For $f\left(\frac{x}{2}\right)$, when $y = {}^-1$, $x = 2$.
For $f({}^-3x)$, when $y = {}^-1$, $x = \frac{{}^-1}{3}$.

Students should notice that for $f(2x)$, $f\left(\frac{x}{2}\right)$ and $f({}^-3x)$, the value of x when $y = {}^-1$ is the reciprocal of k.

Students could then draw a line parallel to the x-axis and compare the values of x where each of the curves crosses the line.

For example, consider the line $y = 8$.

For $f(x)$, when $y = 8$, $x = {}^-2$ and $x = 4$.
For $f(2x)$, when $y = 8$, $x = {}^-1$ and $x = 2$.
For $f\left(\frac{x}{2}\right)$, when $y = 8$, $x = {}^-4$ and $x = 8$.
For $f({}^-3x)$, when $y = 8$, $x = \frac{2}{3}$ and $x = {}^-1\frac{1}{3}$.

Students should notice that for $f(2x)$, $f\left(\frac{x}{2}\right)$ and $f({}^-3x)$, the values of x when $y = 8$ are the values of x when $y = 8$ for $f(x)$ (that is, ${}^-2$ and 4) multiplied by the reciprocal of k.

For $f(2x)$, when $y = 8$, $x = {}^-2 \times \frac{1}{2} = {}^-1$ and $x = 4 \times \frac{1}{2} = 2$.

For $f\left(\frac{x}{2}\right)$, when $y = 8$, $x = {}^-2 \times 2 = {}^-4$ and $x = 4 \times 2 = 8$.

For $f({}^-3x)$, when $y = 8$, $x = {}^-2 \times \frac{{}^-1}{3} = \frac{2}{3}$ and $x = 4 \times \frac{{}^-1}{3} = {}^-1\frac{1}{3}$.

Probability 15

D10.3

Solve problems involving the addition or multiplication of two probabilities.

H4/4g

Objectives

- Solve problems involving the addition or multiplication of two probabilities

Prior knowledge

- Use the addition rule for mutually exclusive events,
  P(A or B) = P(A) + P(B)
- Use the multiplication rule for independent events,
  P(A and B) = P(A) × P(B)
- Draw and use probability tree diagrams
- Find probabilities of dependent events

Equipment needed

None

General notes

This chapter brings together the work on probability done earlier in the course, with the emphasis on problem-solving. Tree diagrams are not used in all the examples and the questions in the exercise do not ask for them; however, unless students have a very clear understanding of the problem, they may find it useful to draw one.

In Example 1 there are two ways of being stopped on just one of the days or, equivalently, two routes through a tree diagram representing the problem. This means that the answer to part **b)** can be found by multiplying the probability of being stopped on one day but not the other by 2.

P(stopped on just one of the days)
= $(0 \cdot 6 \times 0 \cdot 4) + (0 \cdot 4 \times 0 \cdot 6)$
= $2 \times (0 \cdot 6 \times 0 \cdot 4)$

This is a quicker way of finding the answer and may be useful for other, similar questions. When teaching this point, you may find that students see the connection more easily when fractions are used.

P(stopped on just one of the days)
= $\left(\frac{6}{10} \times \frac{4}{10}\right) + \left(\frac{4}{10} \times \frac{6}{10}\right)$
= $2 \times \left(\frac{6}{10} \times \frac{4}{10}\right)$

You can demonstrate that this also works with dependent probabilities.

Example 2 reminds students that they can use the fact P(not A) = 1 – P(A) as a quicker and easier route to some answers.

STAGE
10

Stage 10 Speed-up sheets

SPEED-UP SHEET 1.1

Exercise 1.1 (page 157)

1

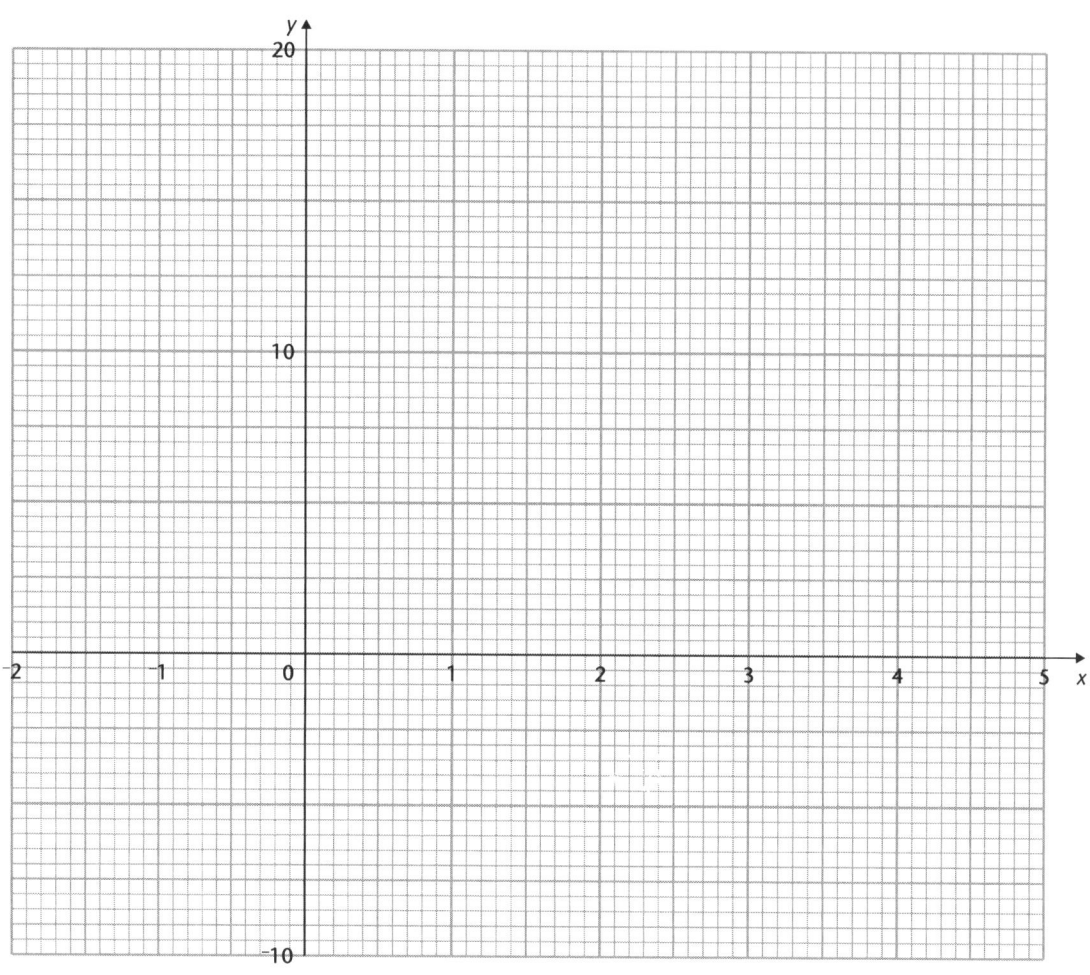

Graduated Assessment for OCR GCSE Mathematics © Hodder Murray 2007

2

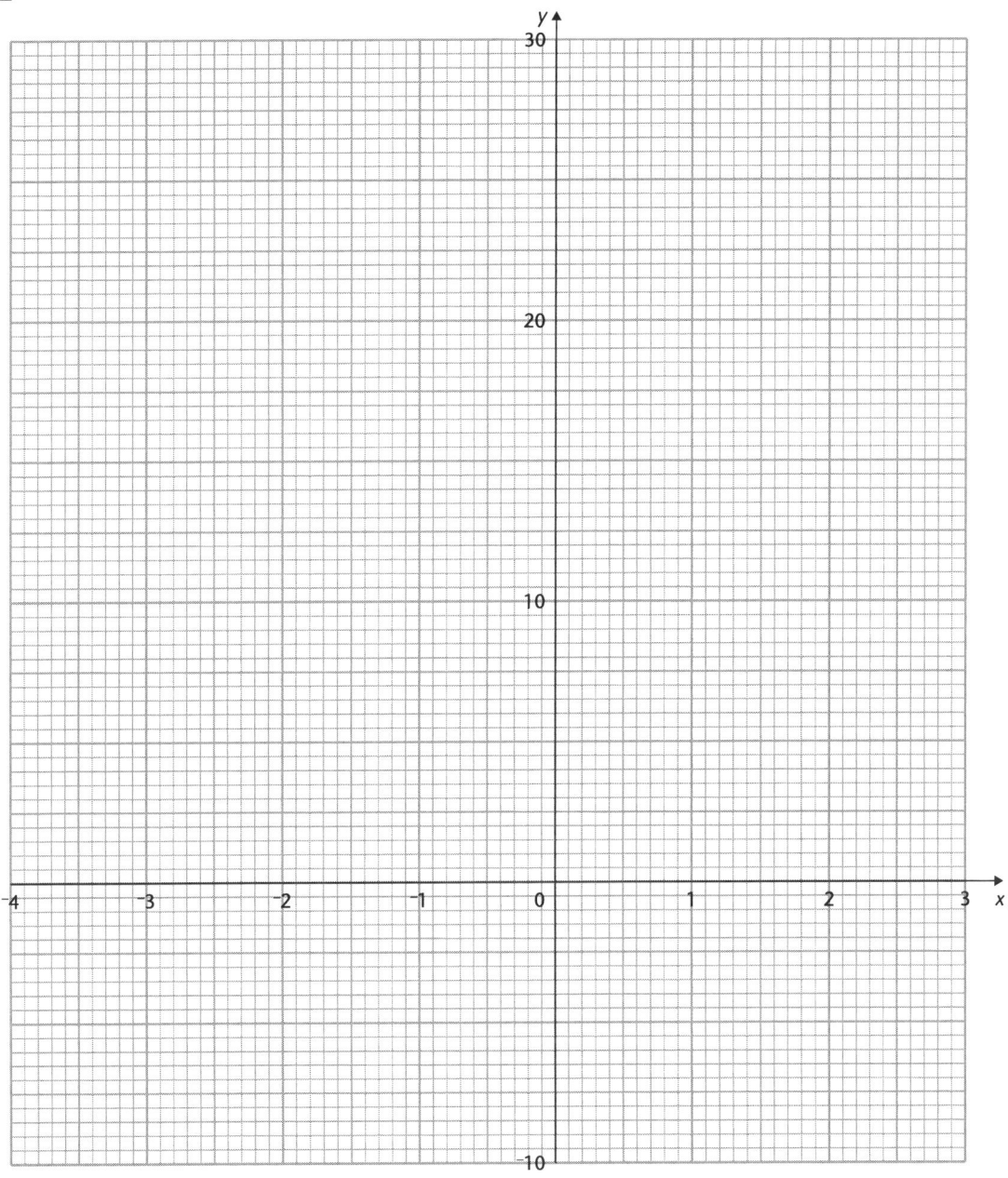

STAGE
10

3

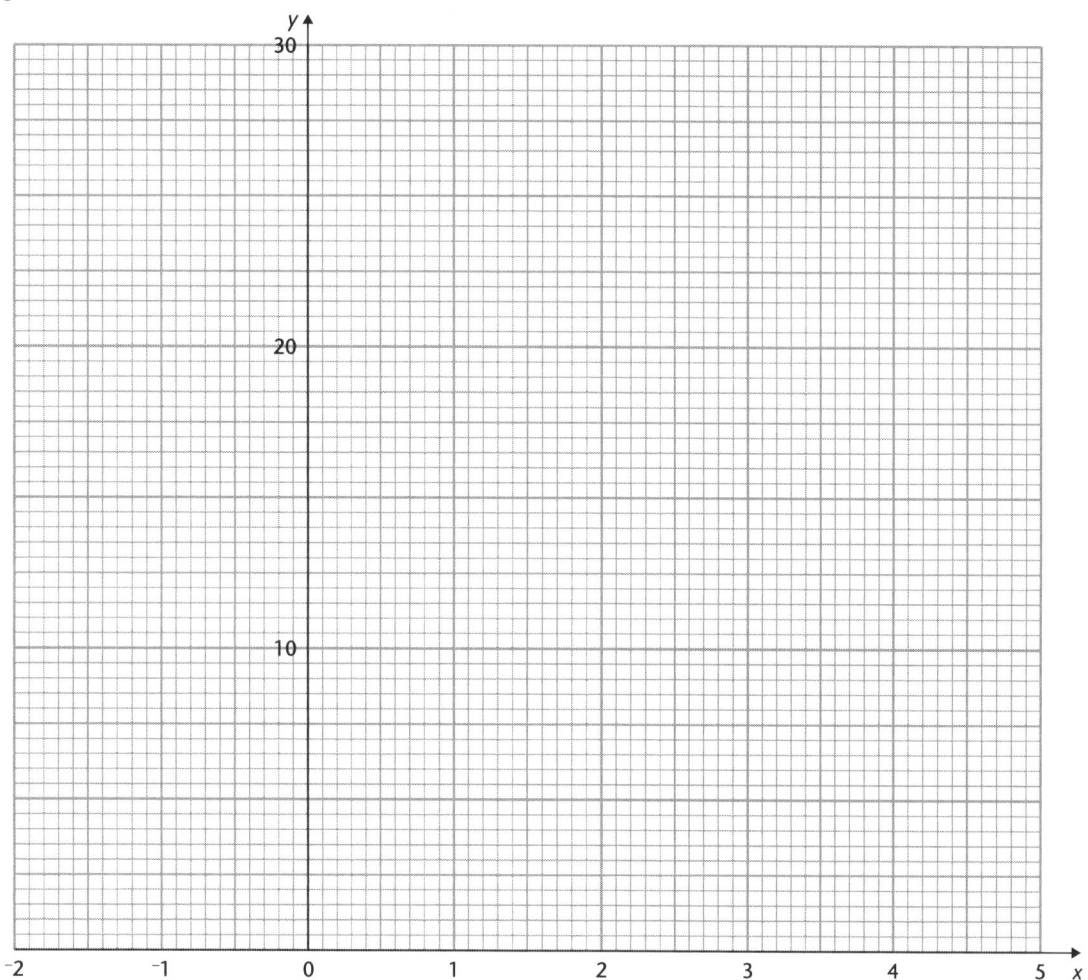

4

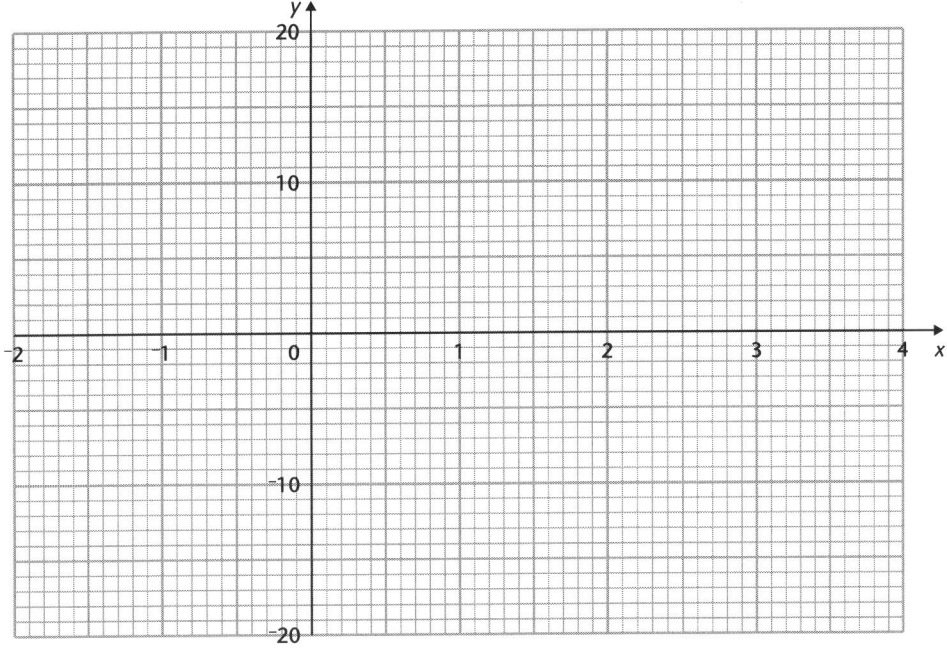

5

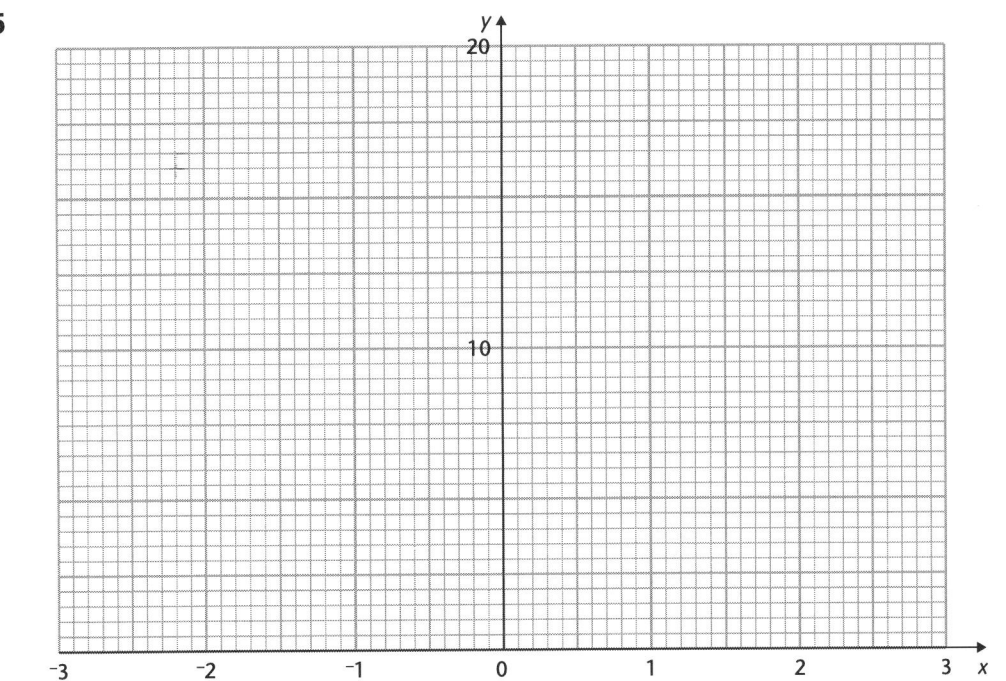

STAGE
10

6

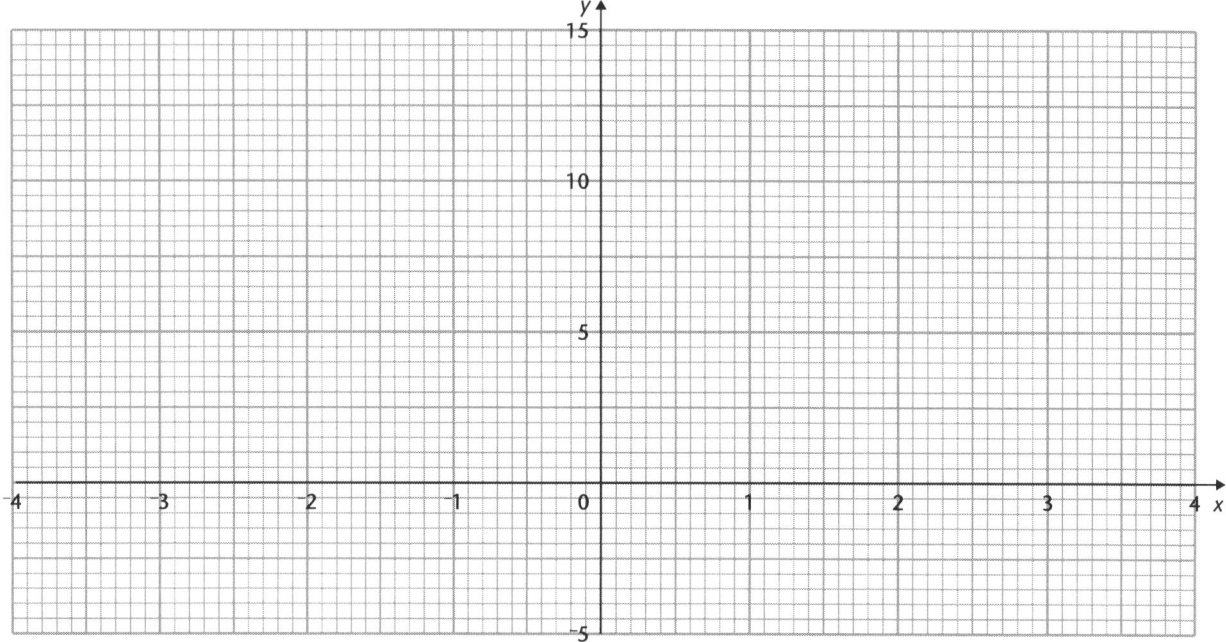

Exercise 1.2 (page 162)

1

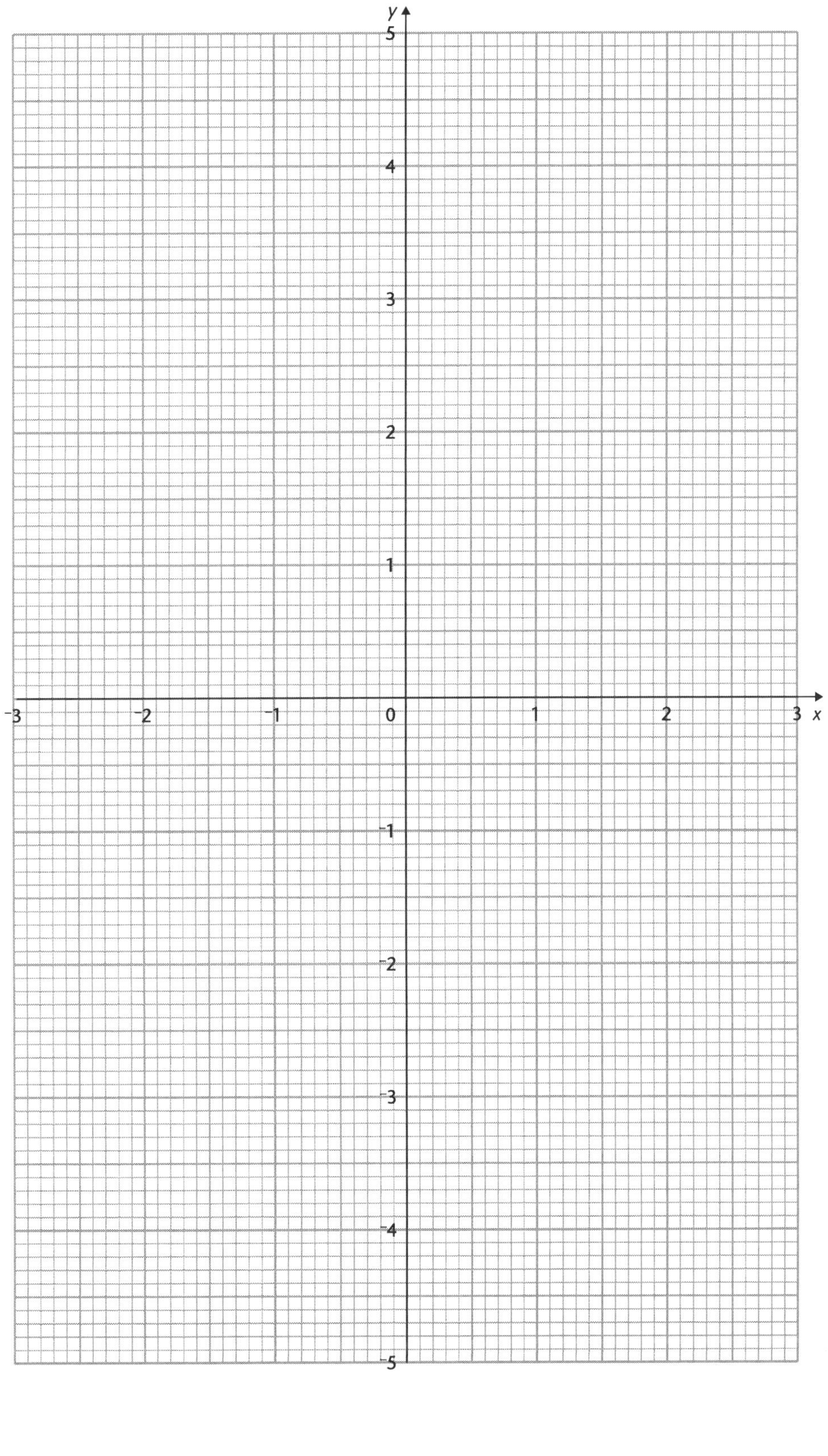

2

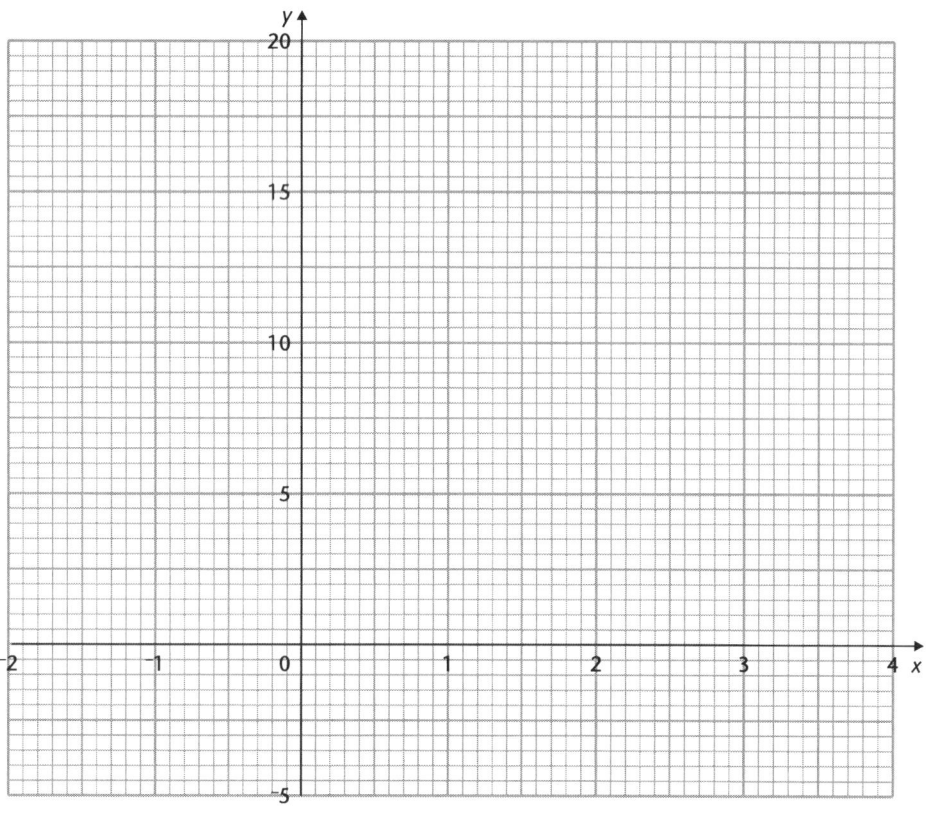

3

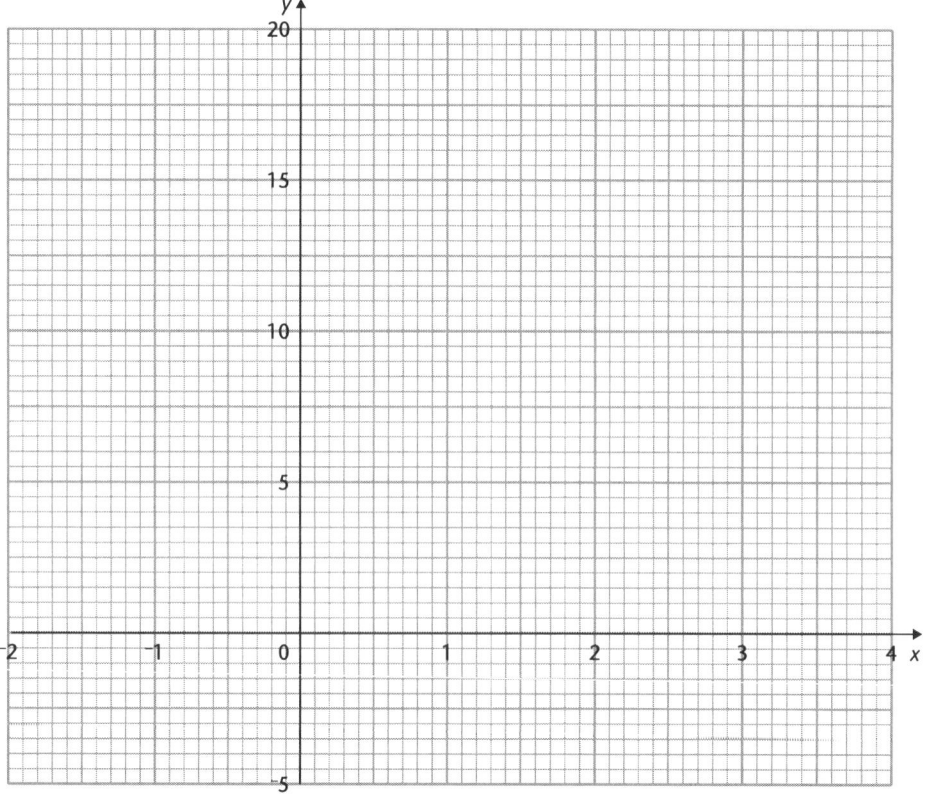

Graduated Assessment for OCR GCSE Mathematics © Hodder Murray 2007

Speed-up sheets

4

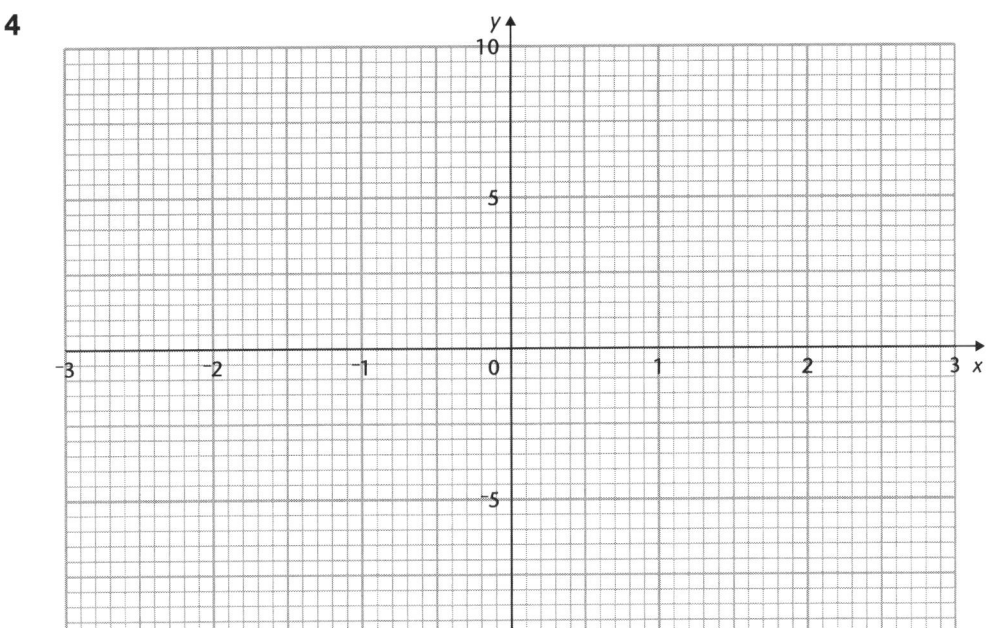

5

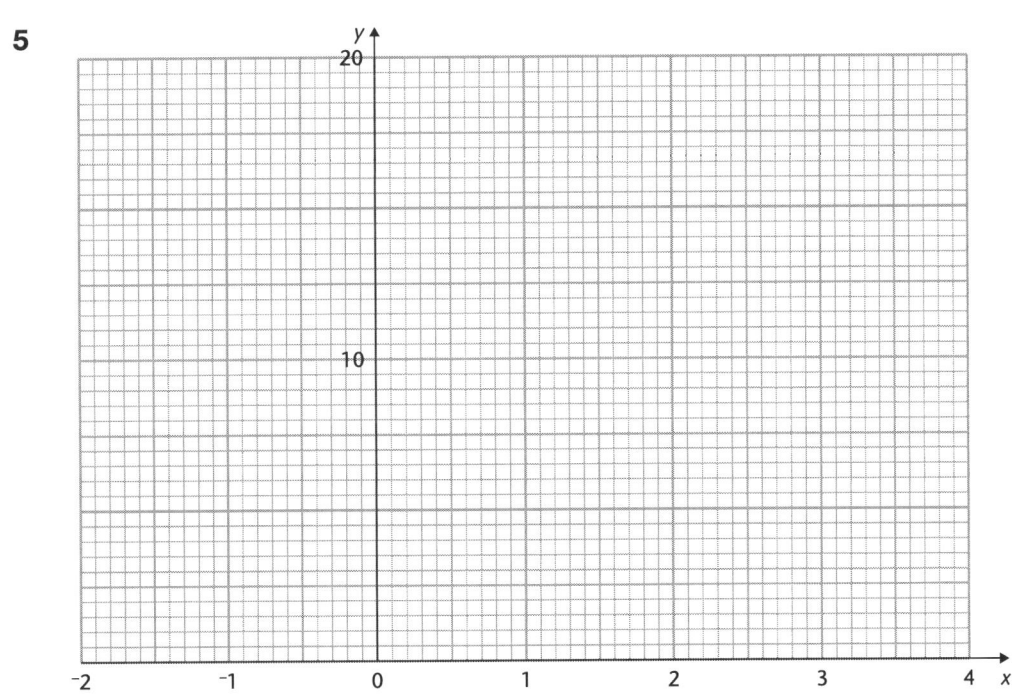

STAGE
10

6

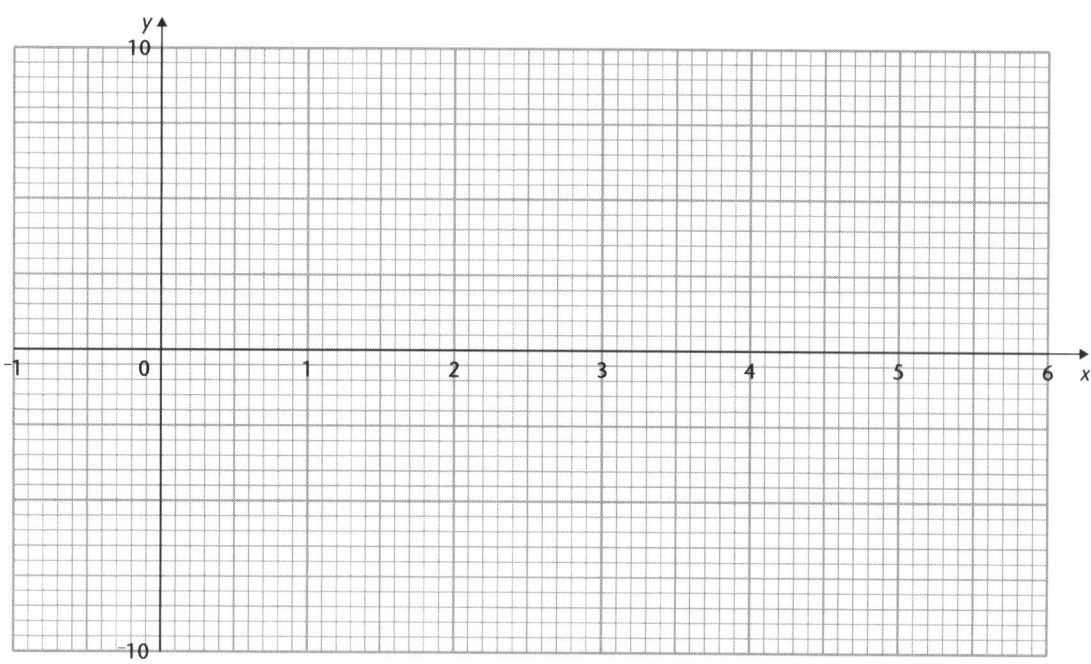

7

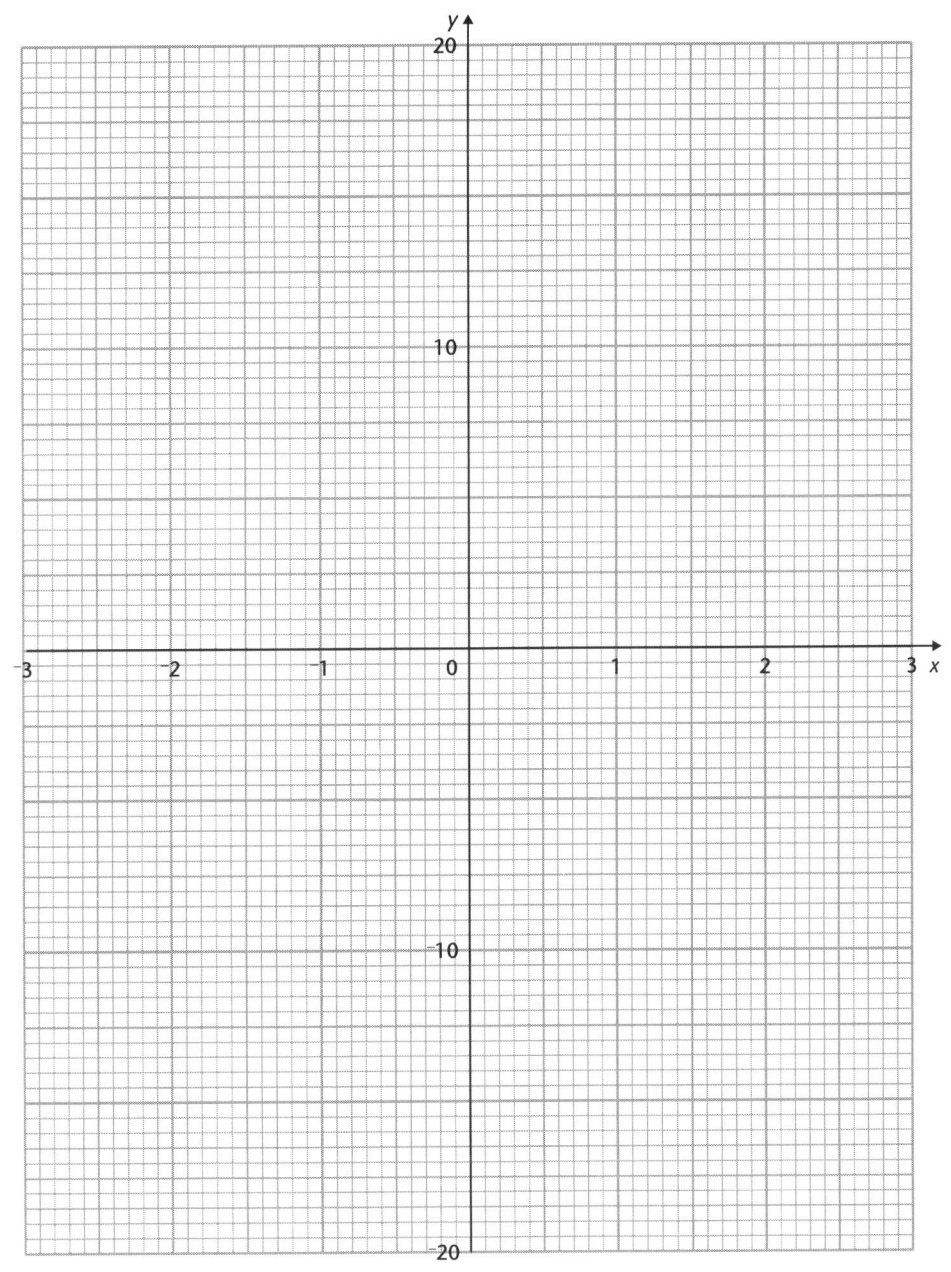

STAGE
10

8

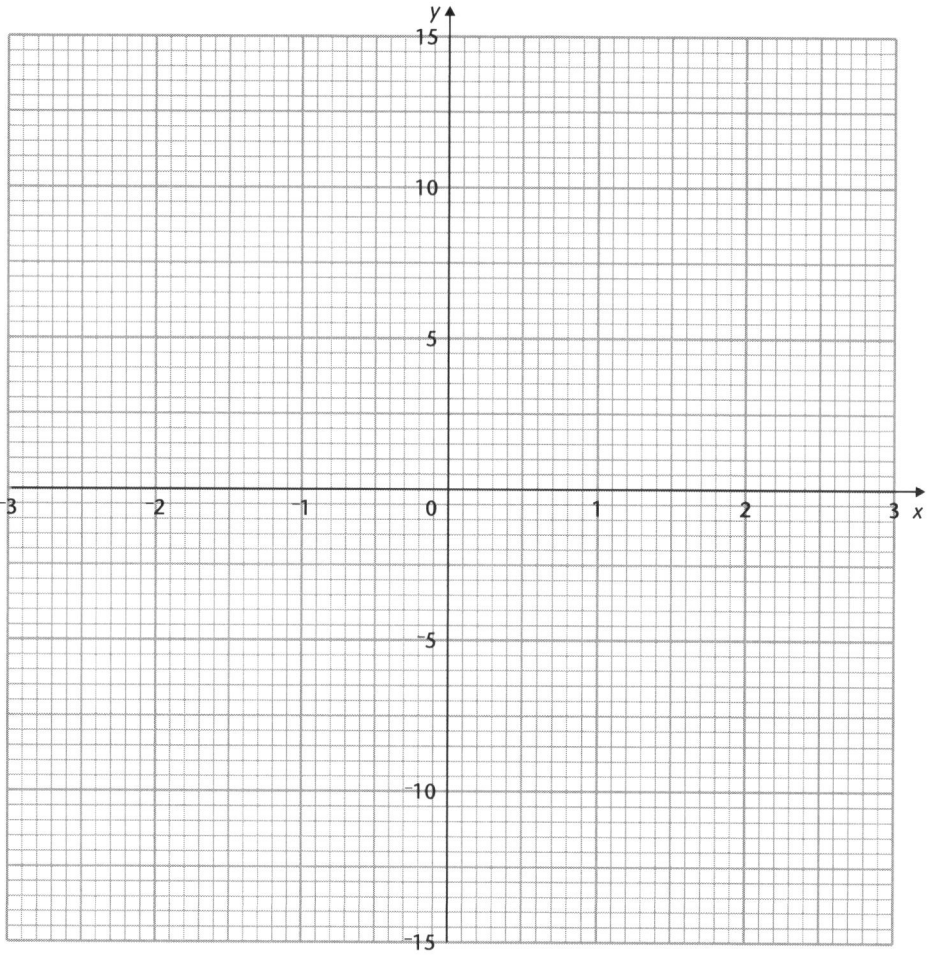

Graduated Assessment for OCR GCSE Mathematics © Hodder Murray 2007

9

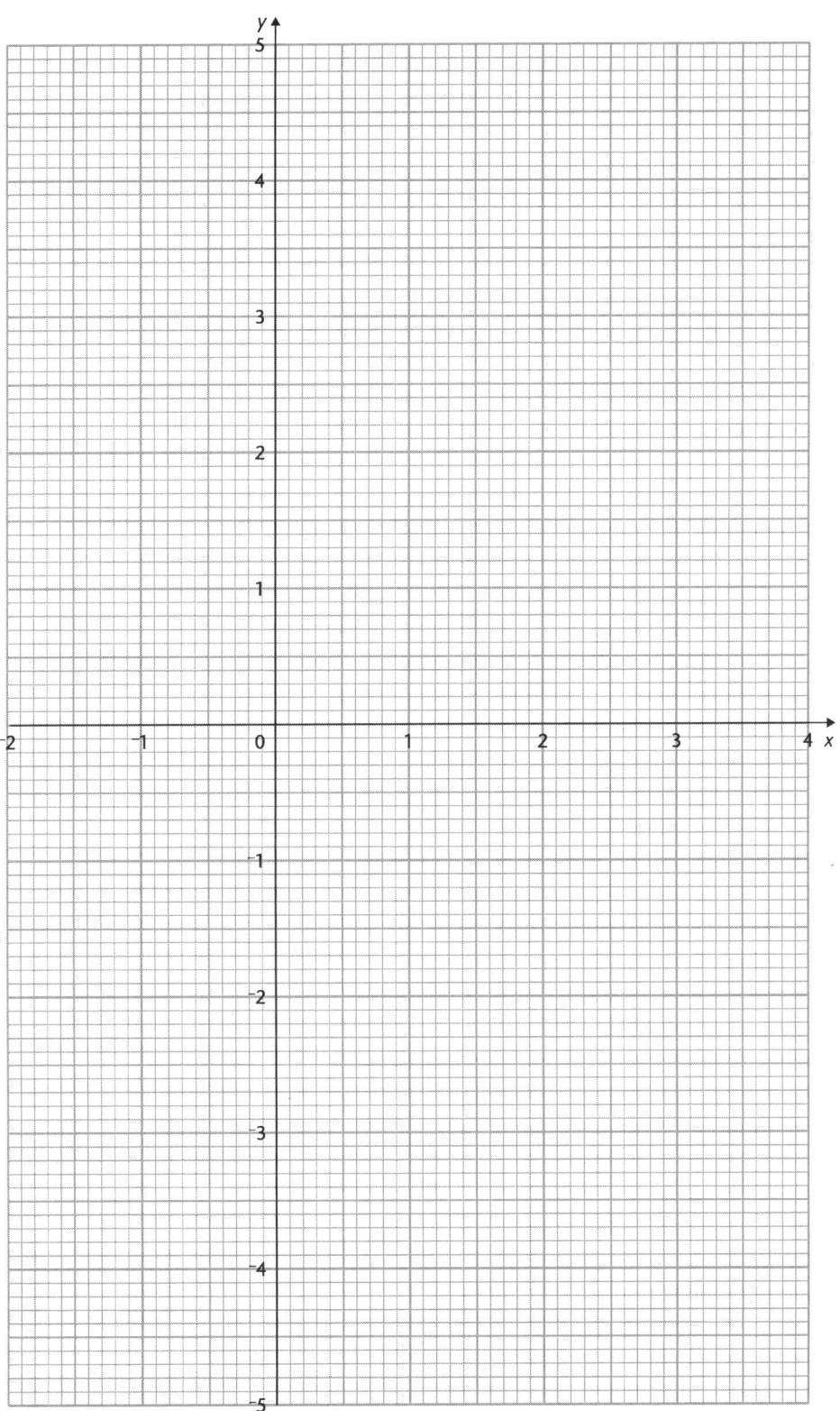

STAGE
10

10

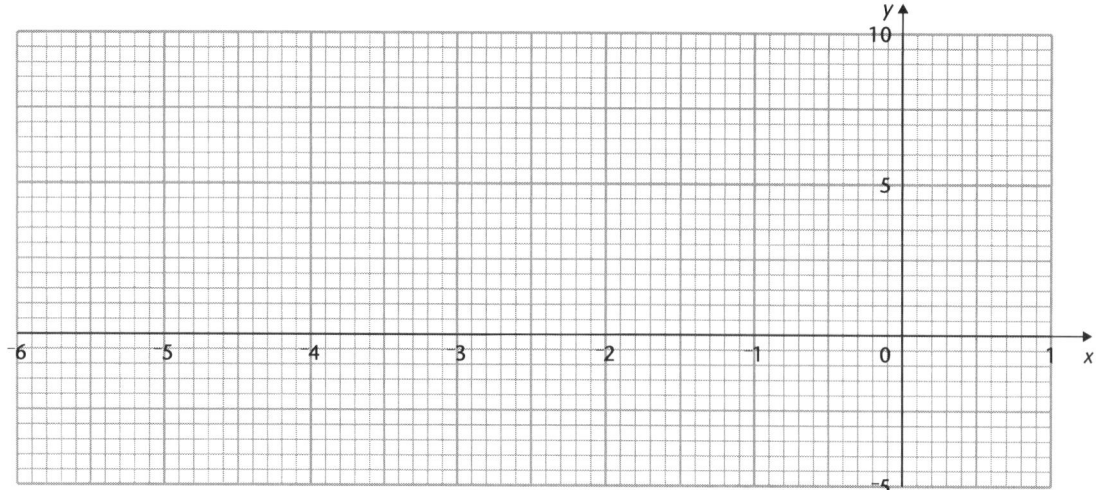

Exercise 1.3 (page 166)

1

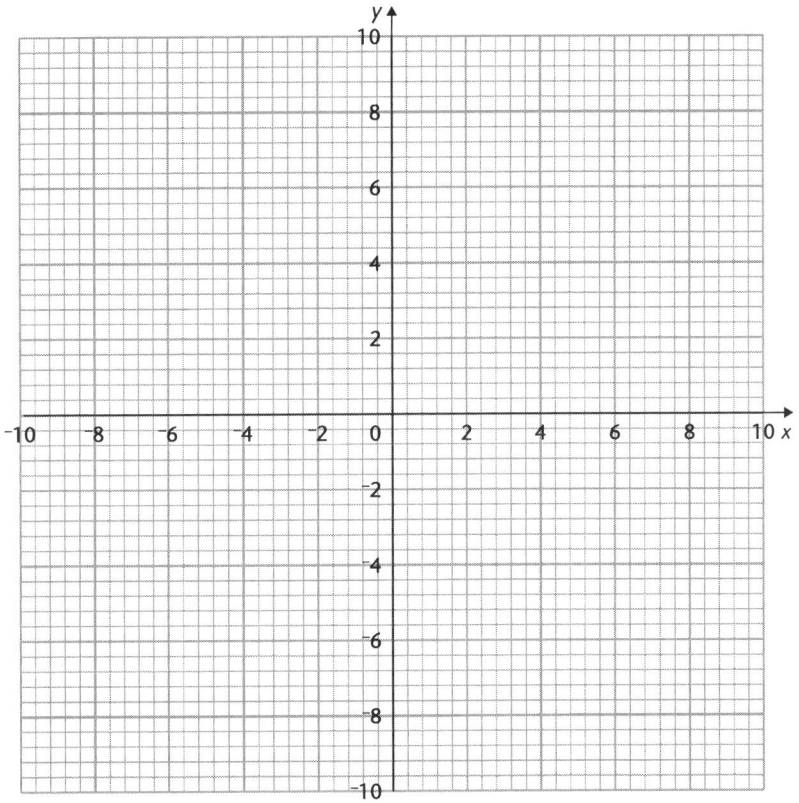

2

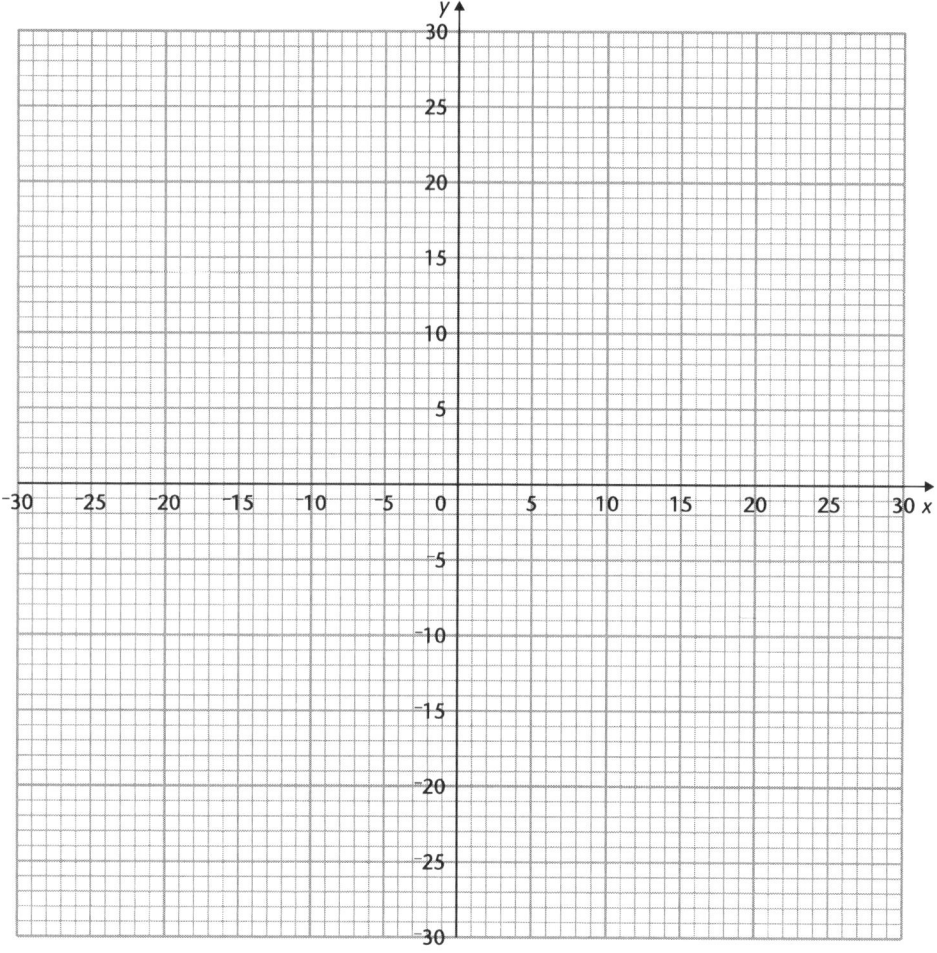

3

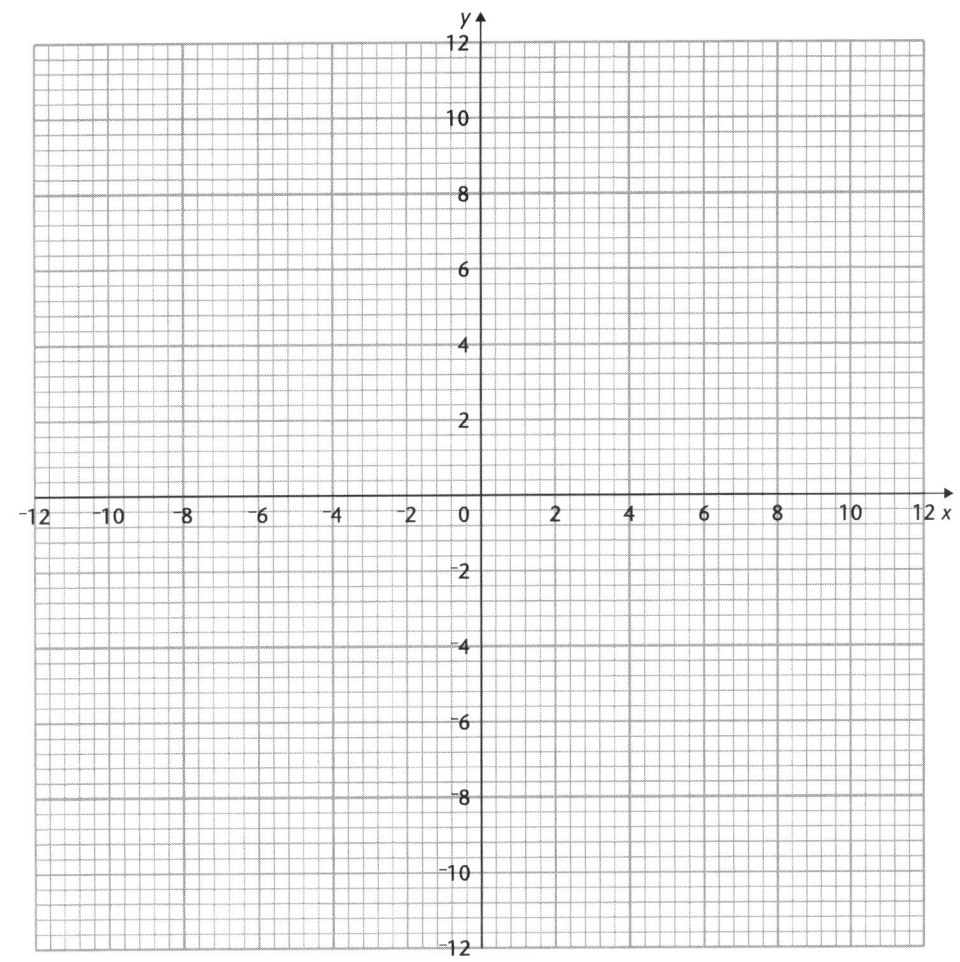

STAGE
10

4

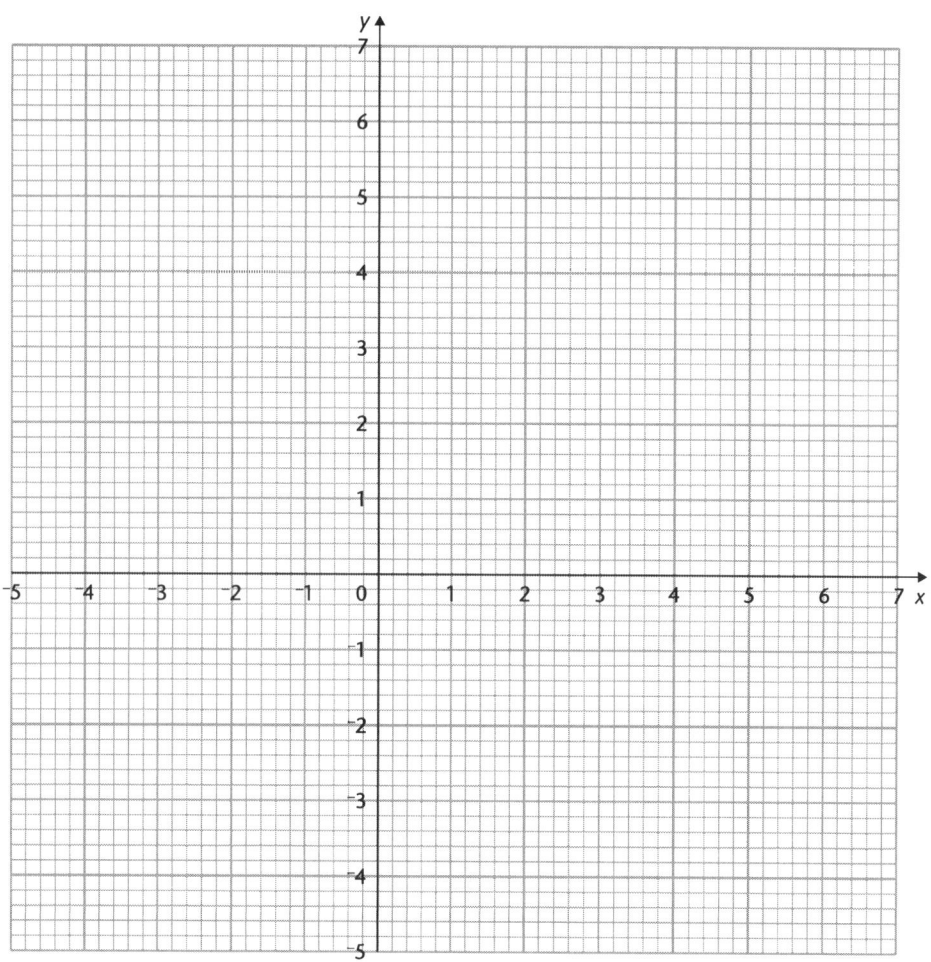

5

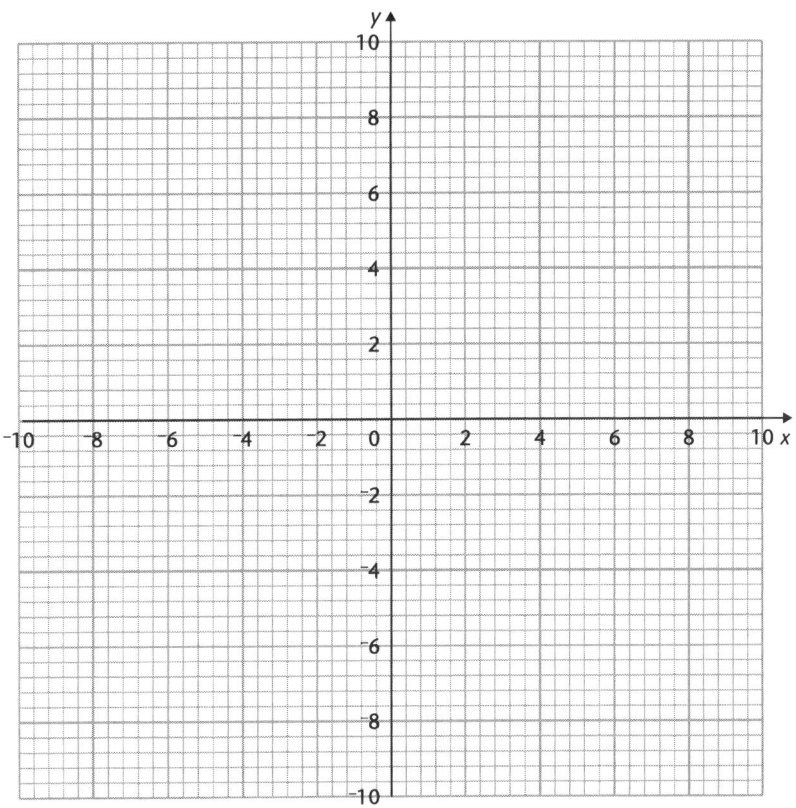

6

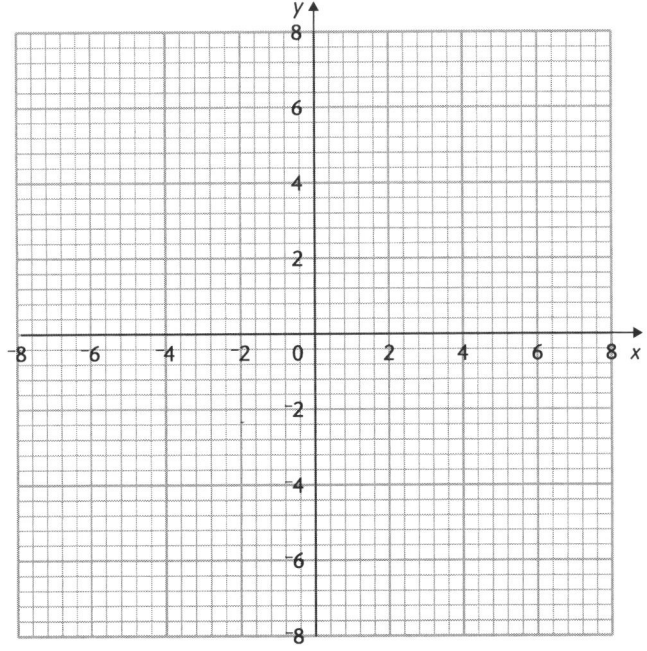

STAGE
10

Speed-up sheets

SPEED-UP SHEET 2.1

Exercise 2.1 (page 174)

3

x	$^-2$	$^-1$	0	1	2	3	4	5
$y = 2^x$								

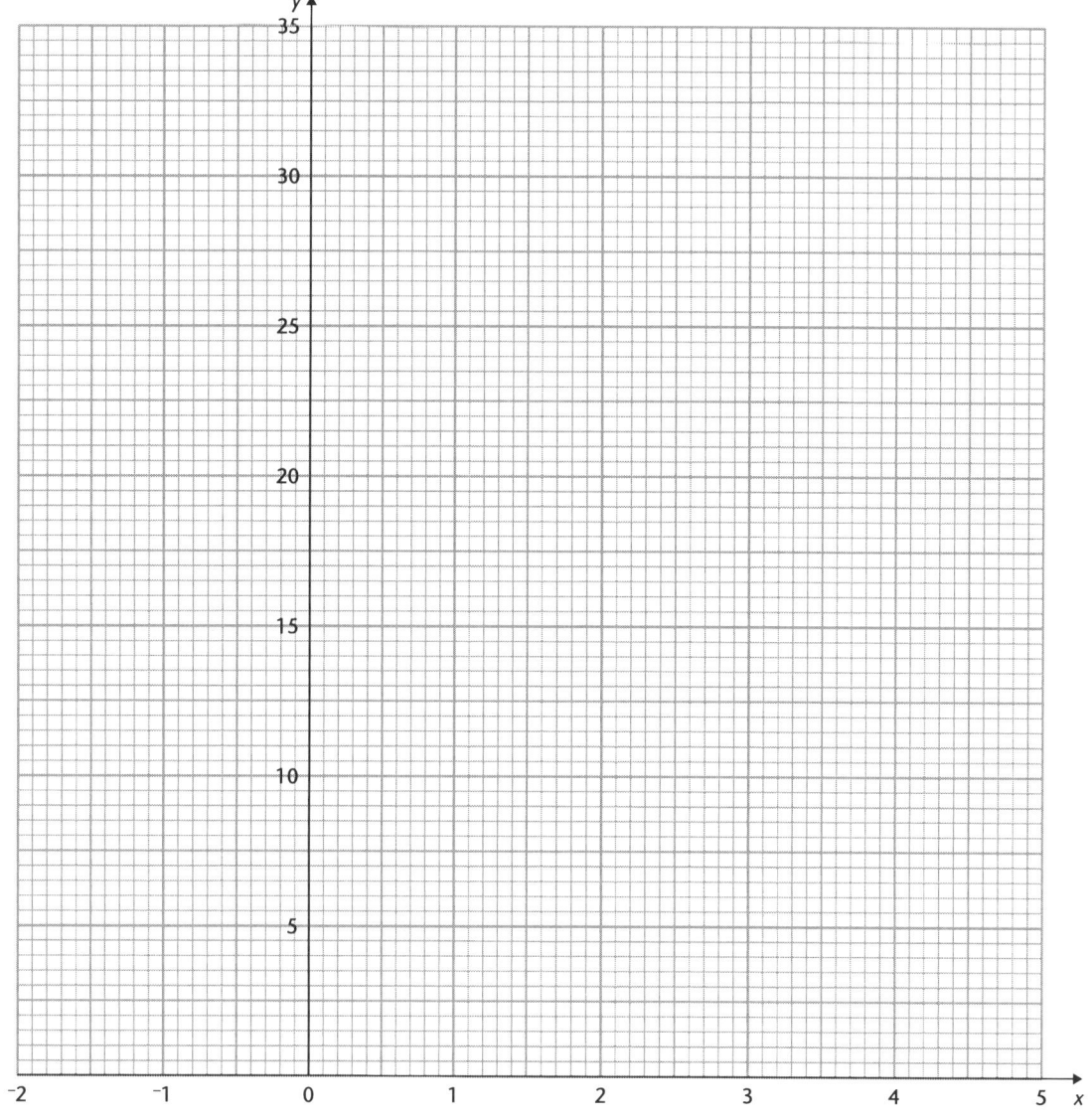

4

x	⁻3	⁻2	⁻1	0	1	2	3	4	5
$y = 1 \cdot 5^x$									

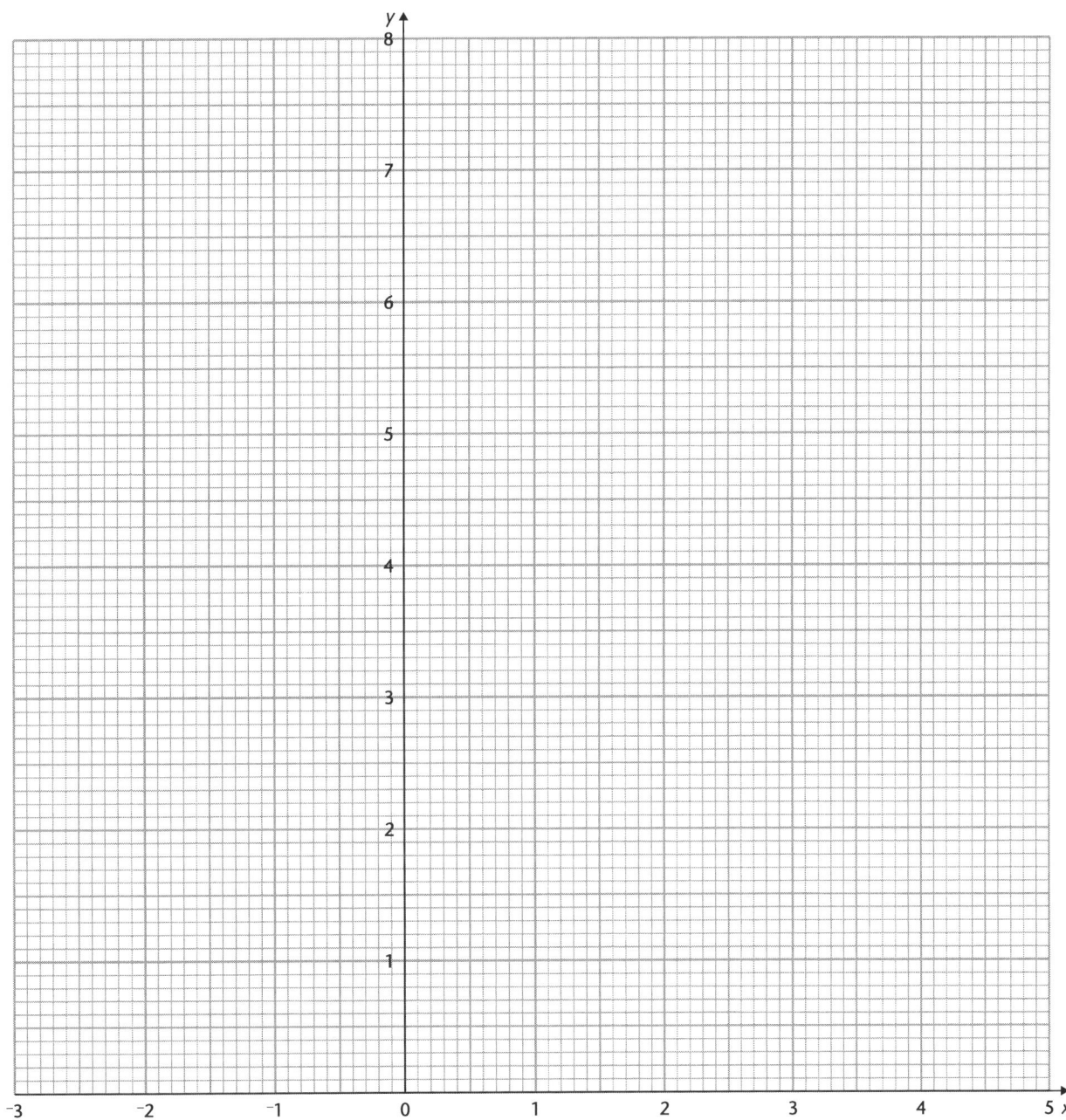

STAGE
10

Speed-up sheets

5

x	0	0·5	1	1·5	2	2·5	3	3·5	4
$y = 3^{-x}$									

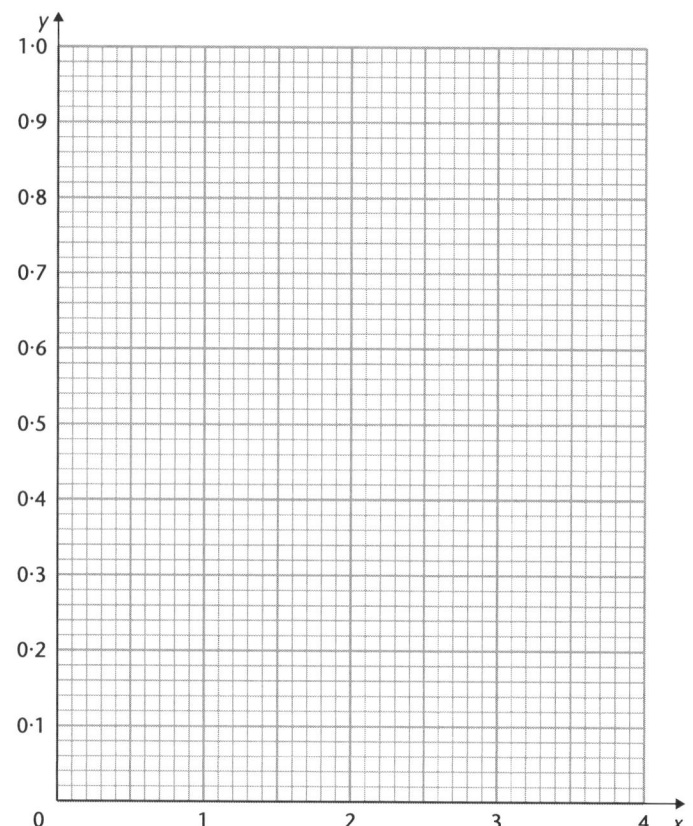

6

x	-2·5	-2	-1·5	-1	-0·5	0	0·5	1
$y = 4^{-x}$								

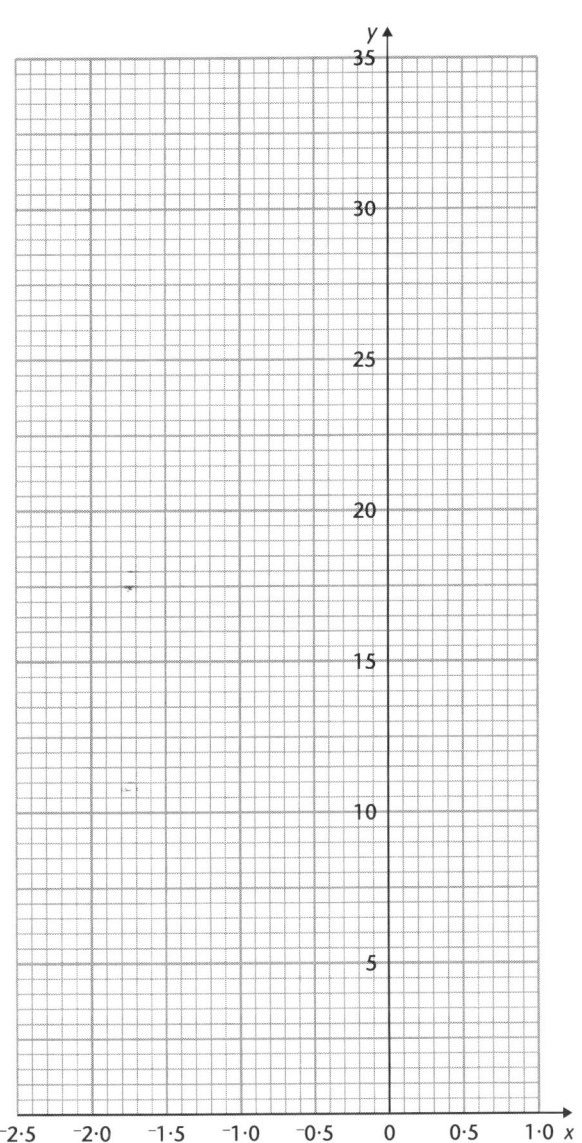

STAGE
10

Speed-up sheets

SPEED-UP SHEET 2.2

Exercise 2.2 (page 177)

1

x	y
0	500
1	560
2	627·20
3	
4	
5	

2

x	y = 3^x
0	
1	
2	
3	
4	
5	

4

Time (hours)	Number of bacteria
0·00	
0·25	
0·50	
0·75	
1·00	
1·25	
1·50	
1·75	
2·00	

Graduated Assessment for OCR GCSE Mathematics © Hodder Murray 2007

Revision exercise A1 (page 207)

1

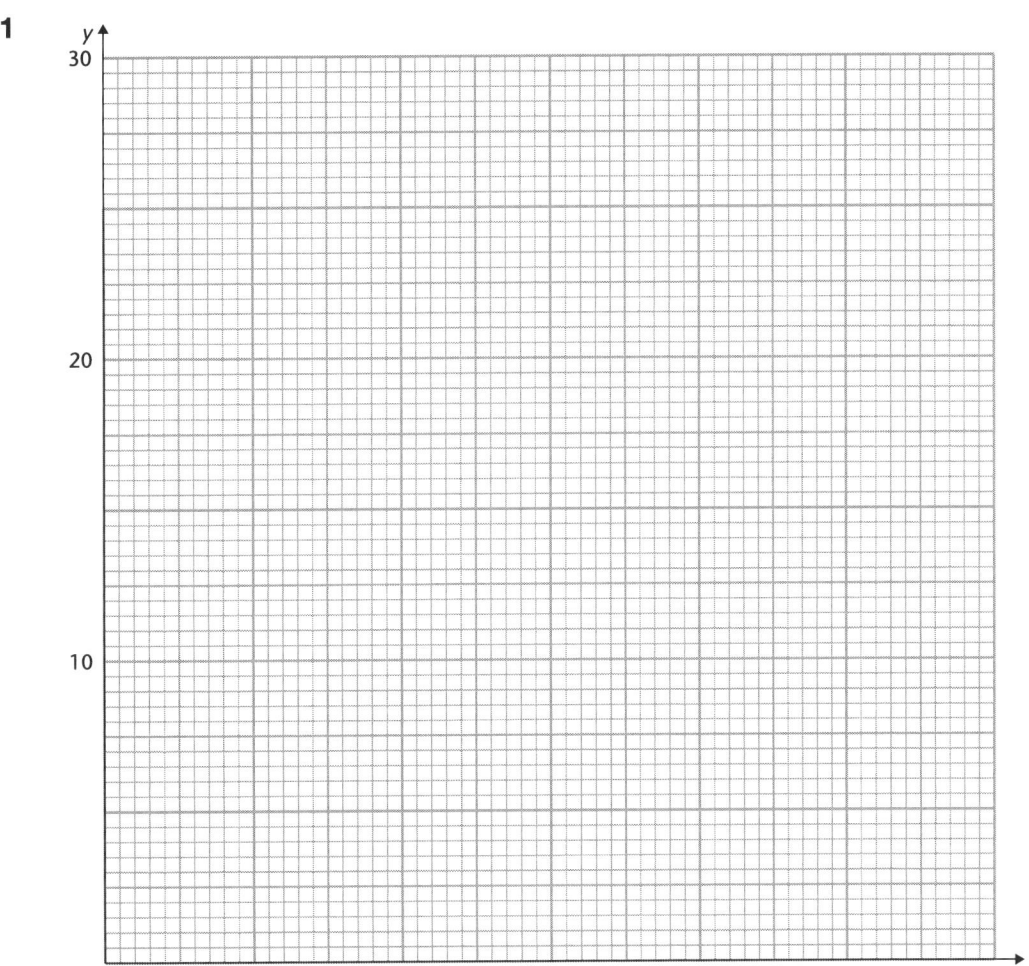

2

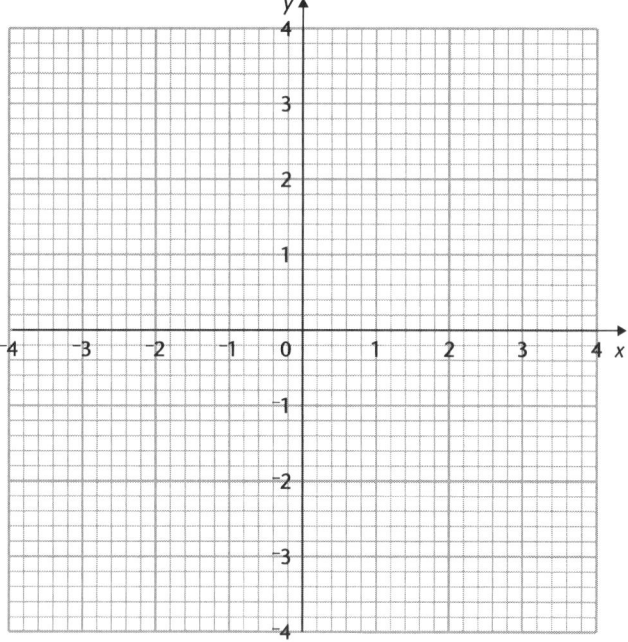

Graduated Assessment for OCR GCSE Mathematics © Hodder Murray 2007

3

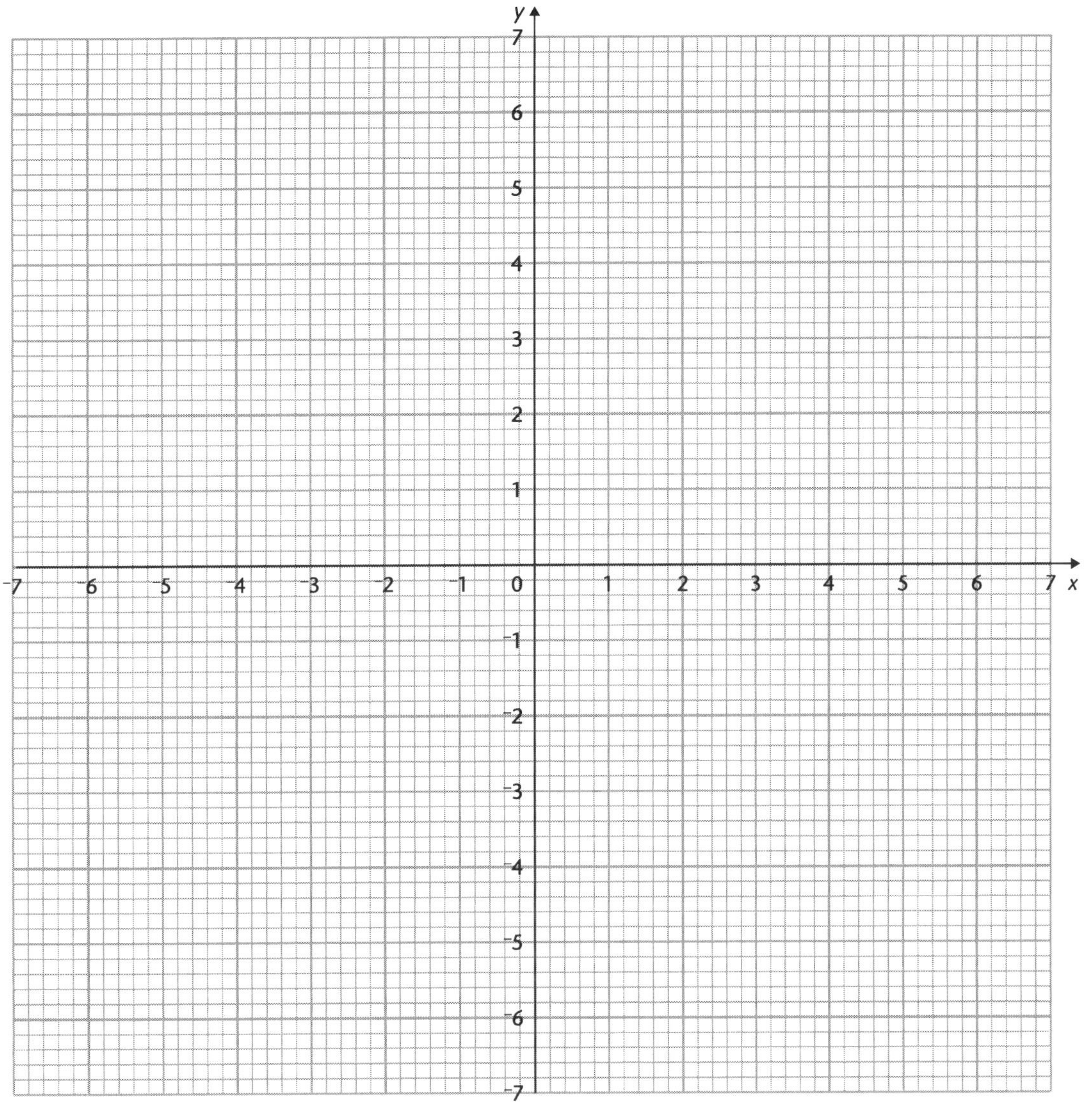

STAGE
10

4

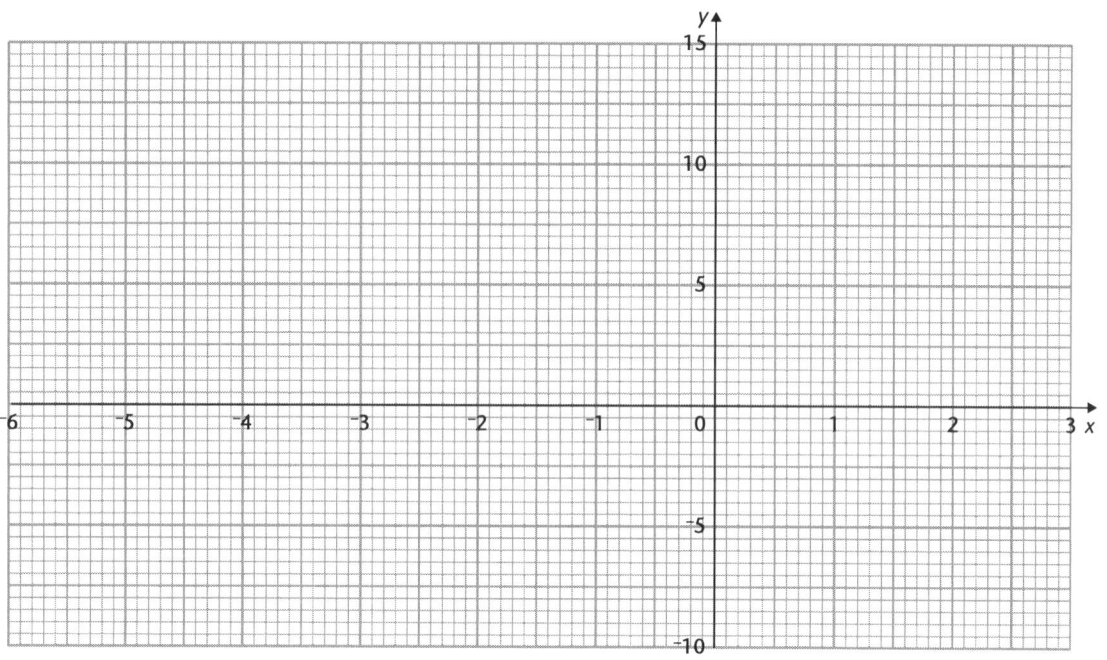

6

x	0	0·5	1	1·5	2	2·5	3	3·5	4
$y = 2^{-x}$	1								

Exercise 5.1 (page 213)

1

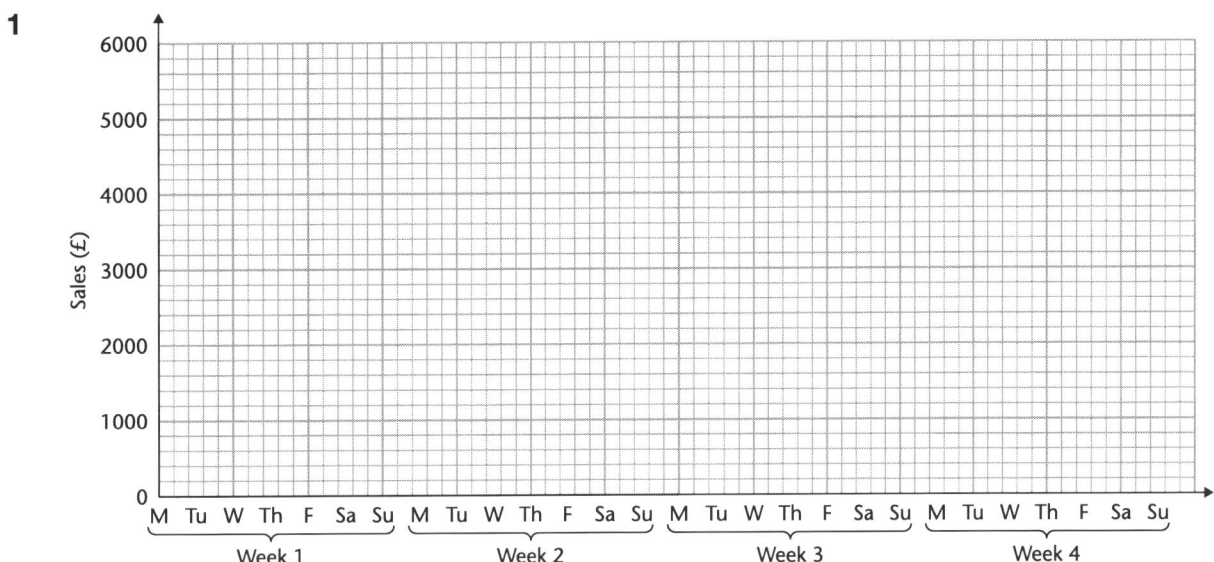

3

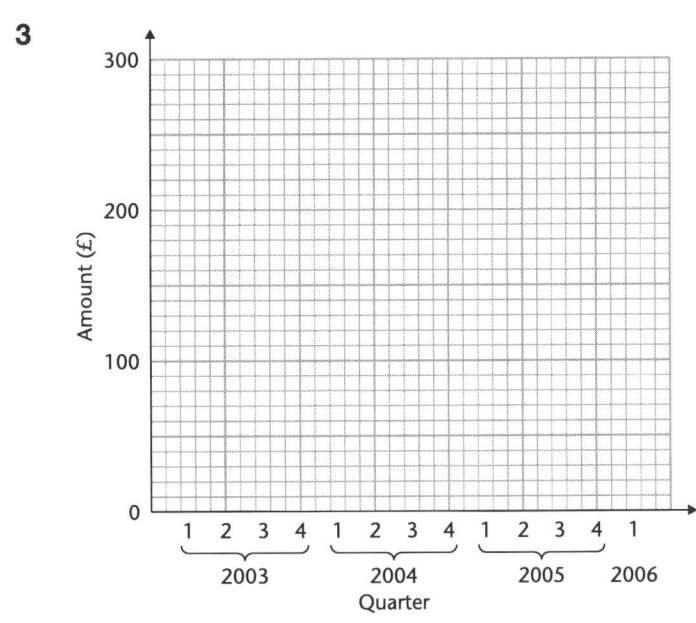

STAGE
10

4

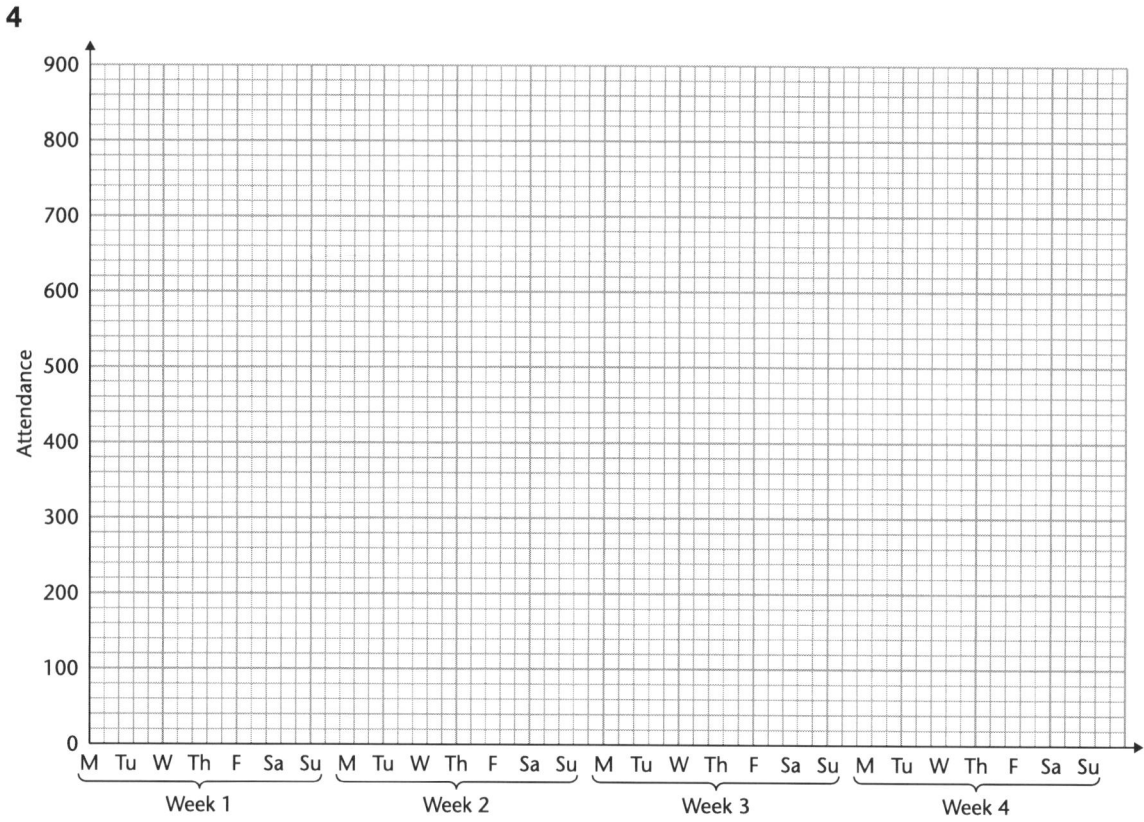

6

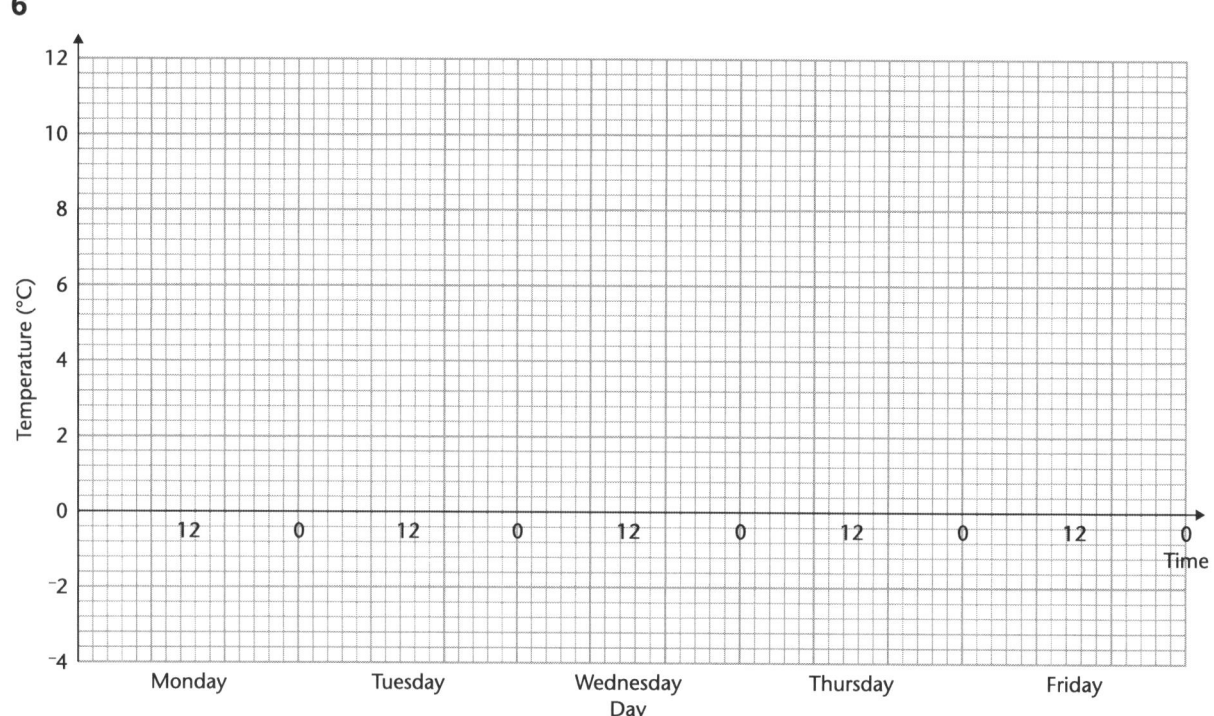

11

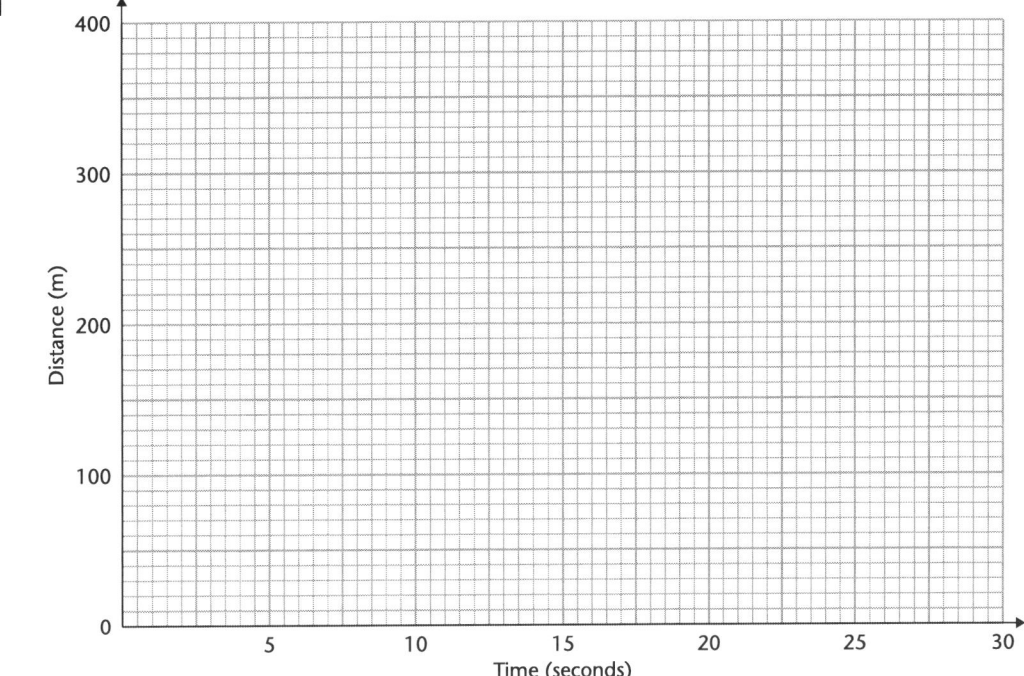

SPEED-UP SHEET B1

Revision exercise B1 (page 246)

1

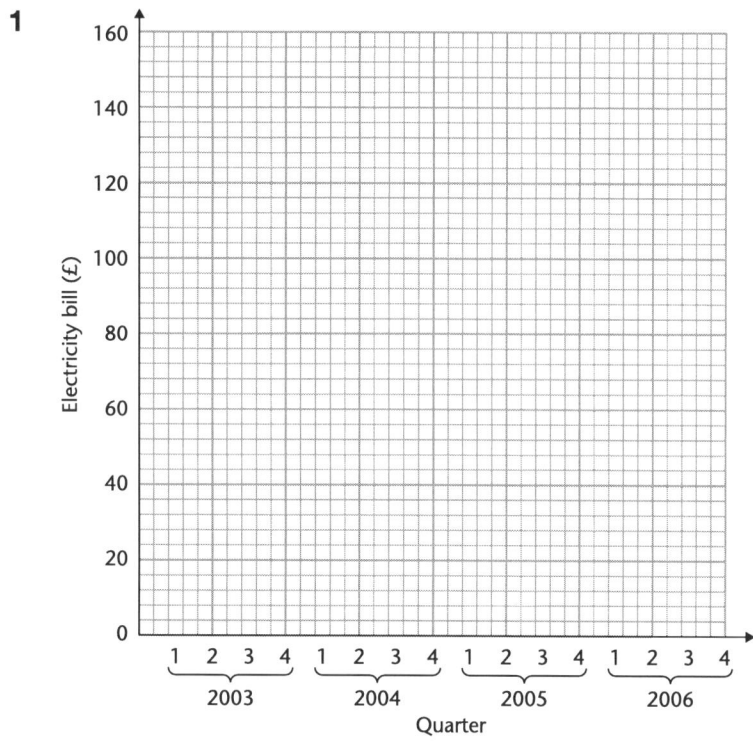

2

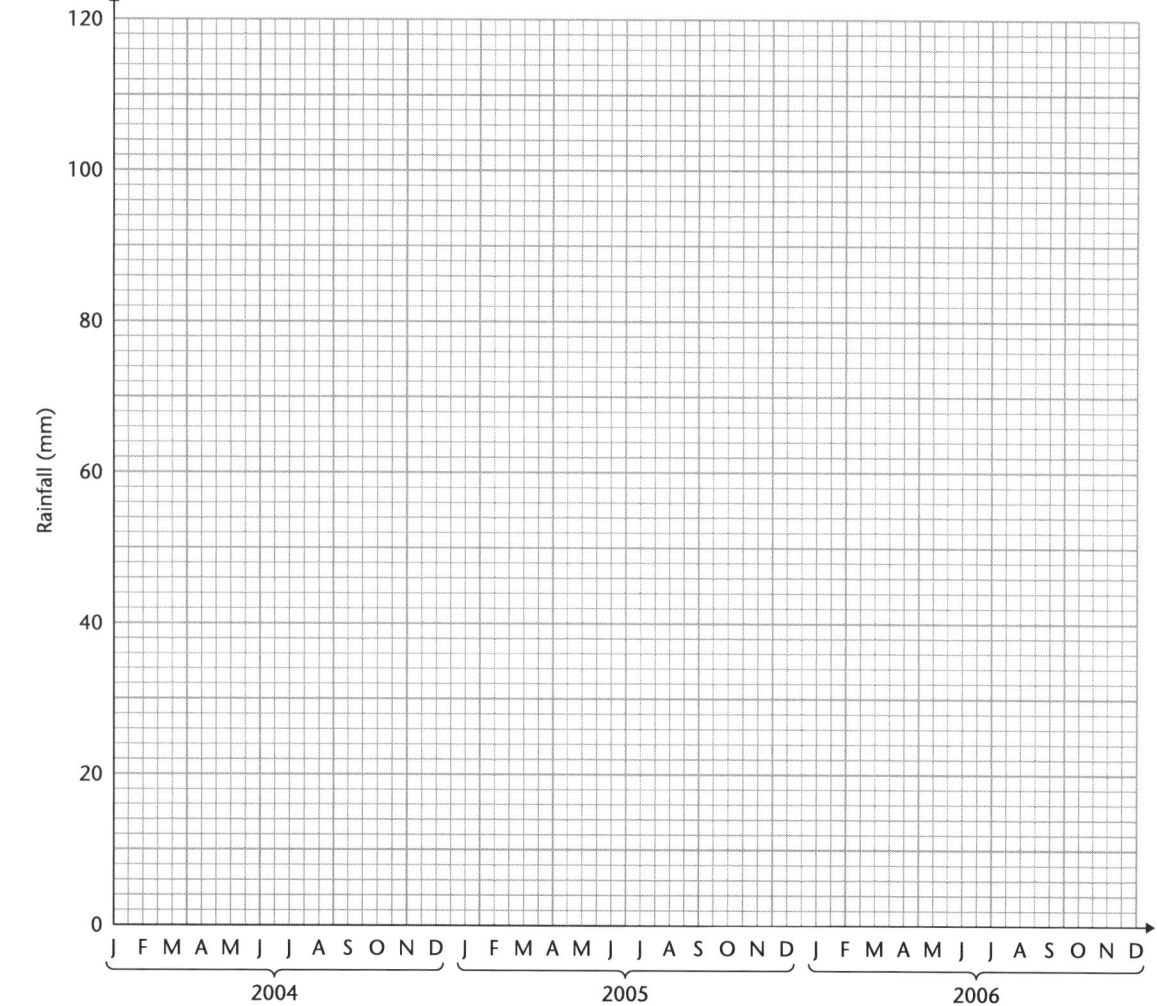

Graduated Assessment for OCR GCSE Mathematics © Hodder Murray 2007

Exercise 10.3 (page 269)

1
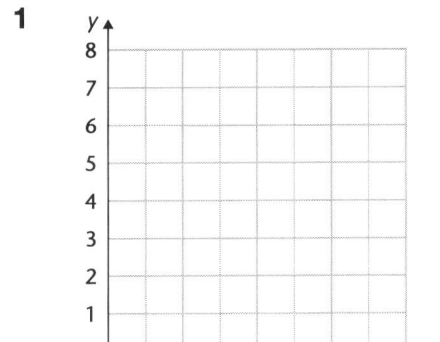

2
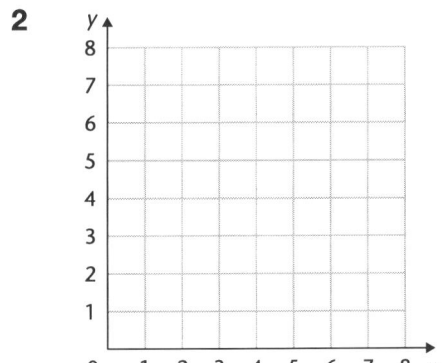

3
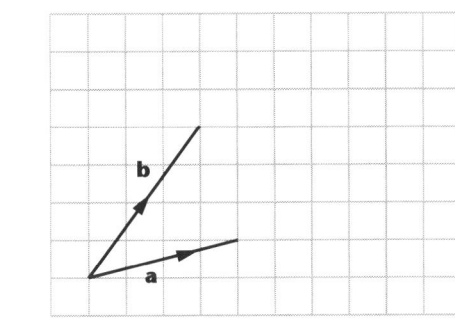

4
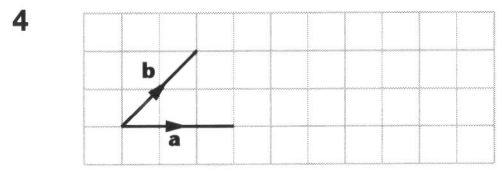

SPEED-UP SHEET 11.1

Exercise 11.1 (page 278)

1

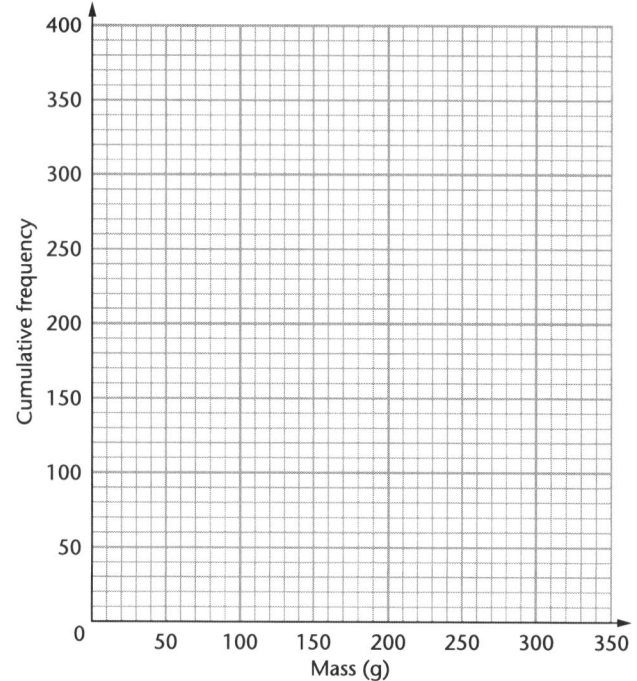

2

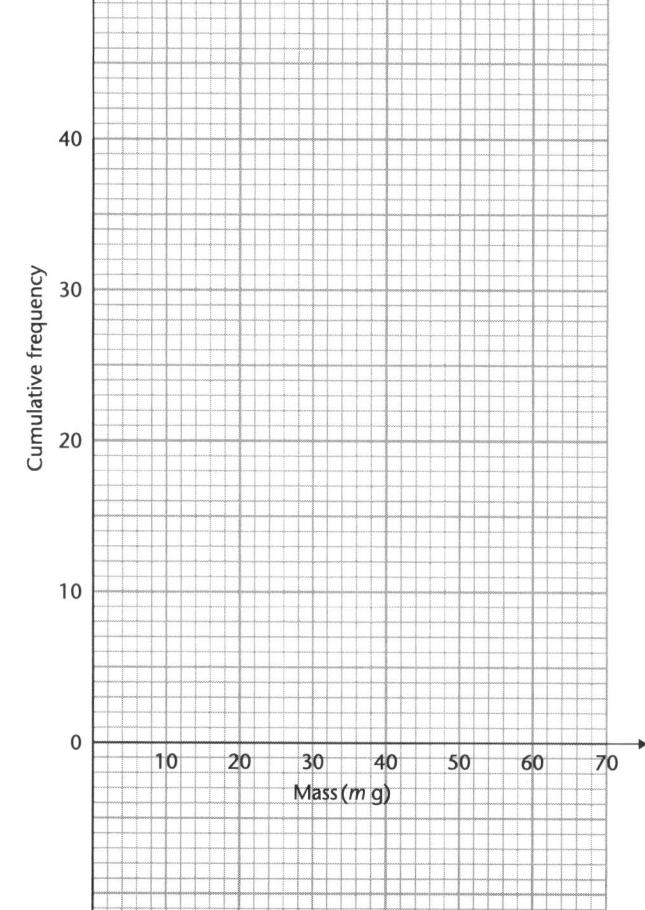

4

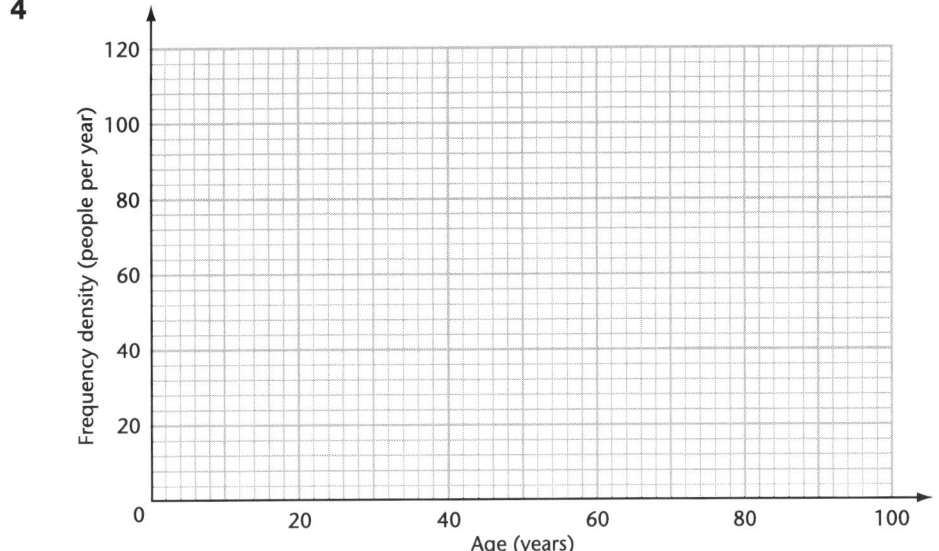

5

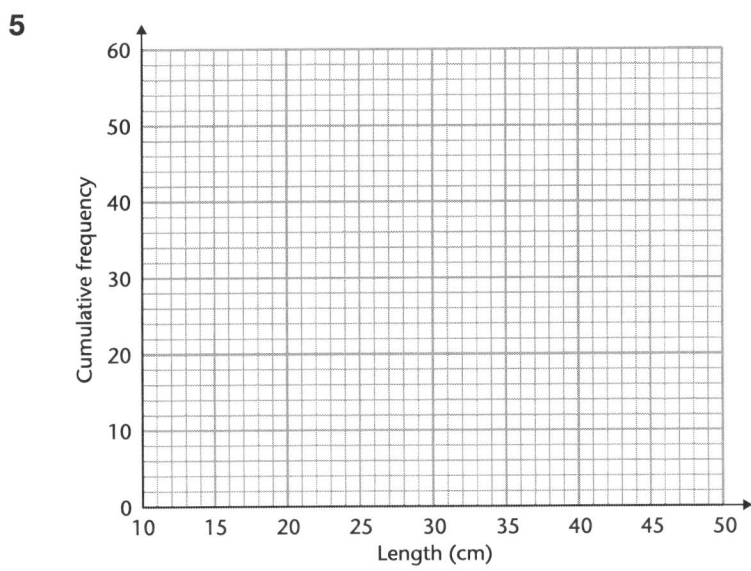

STAGE
10

13

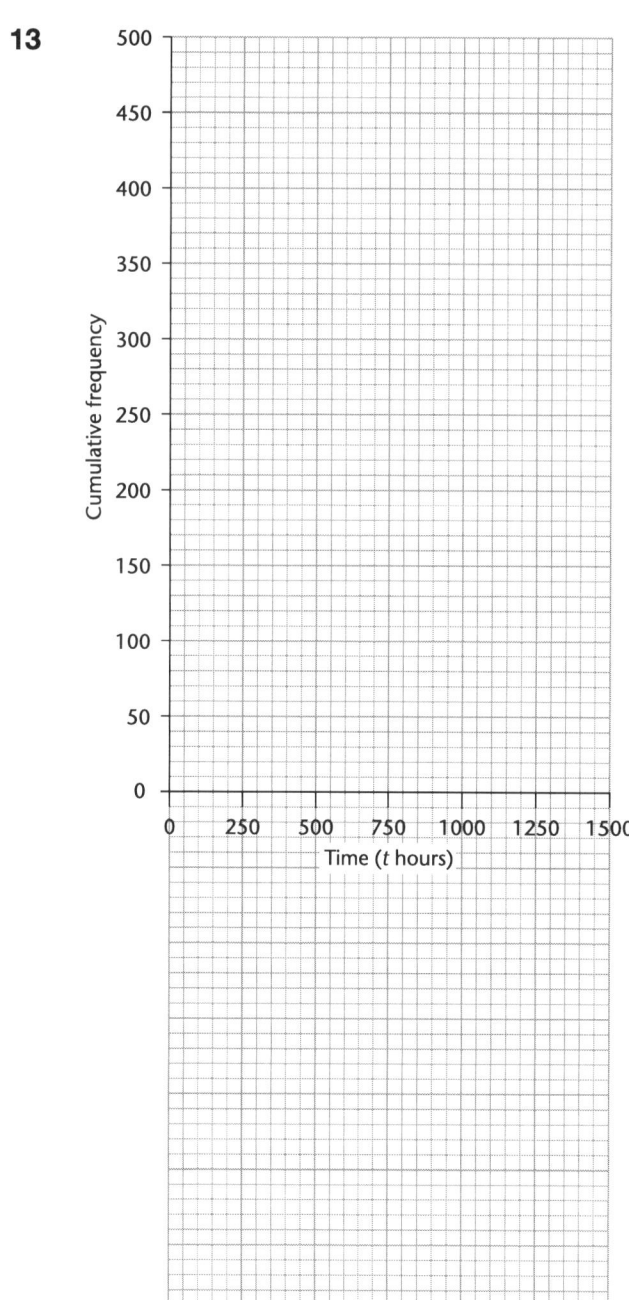

15

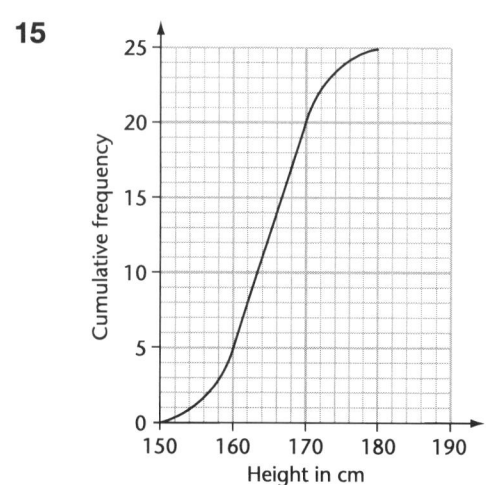

Revision exercise C1 (page 297)

11

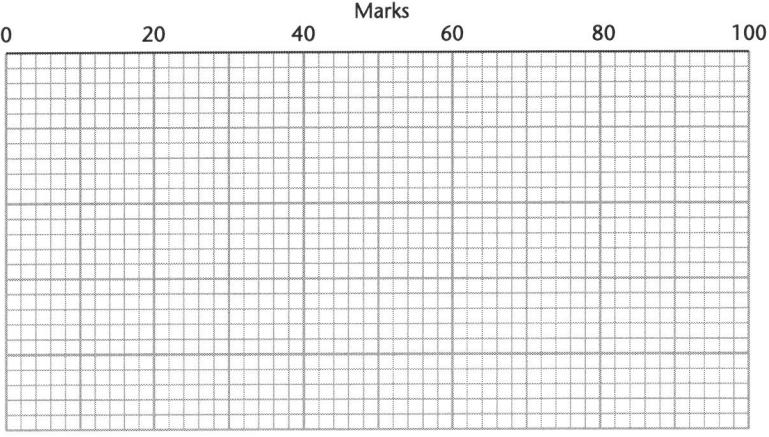

12

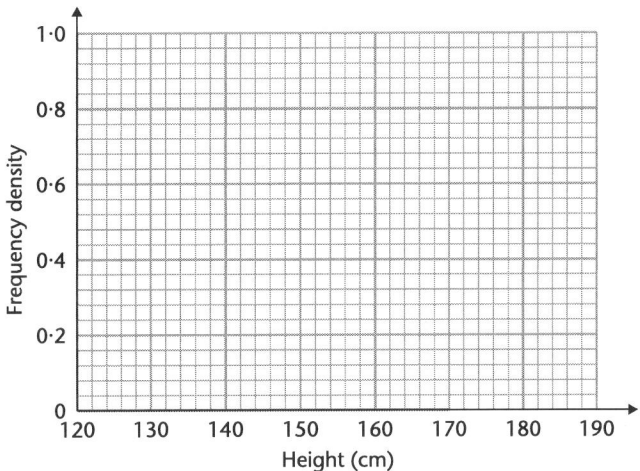

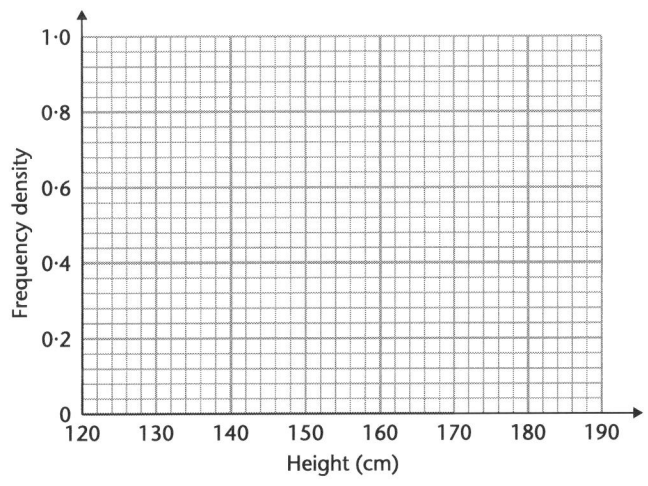

STAGE
10

Speed-up sheets

Exercise 13.1 (page 304)

1

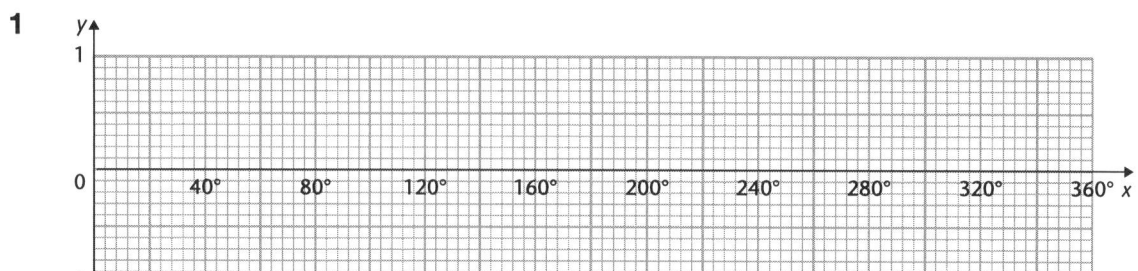

2

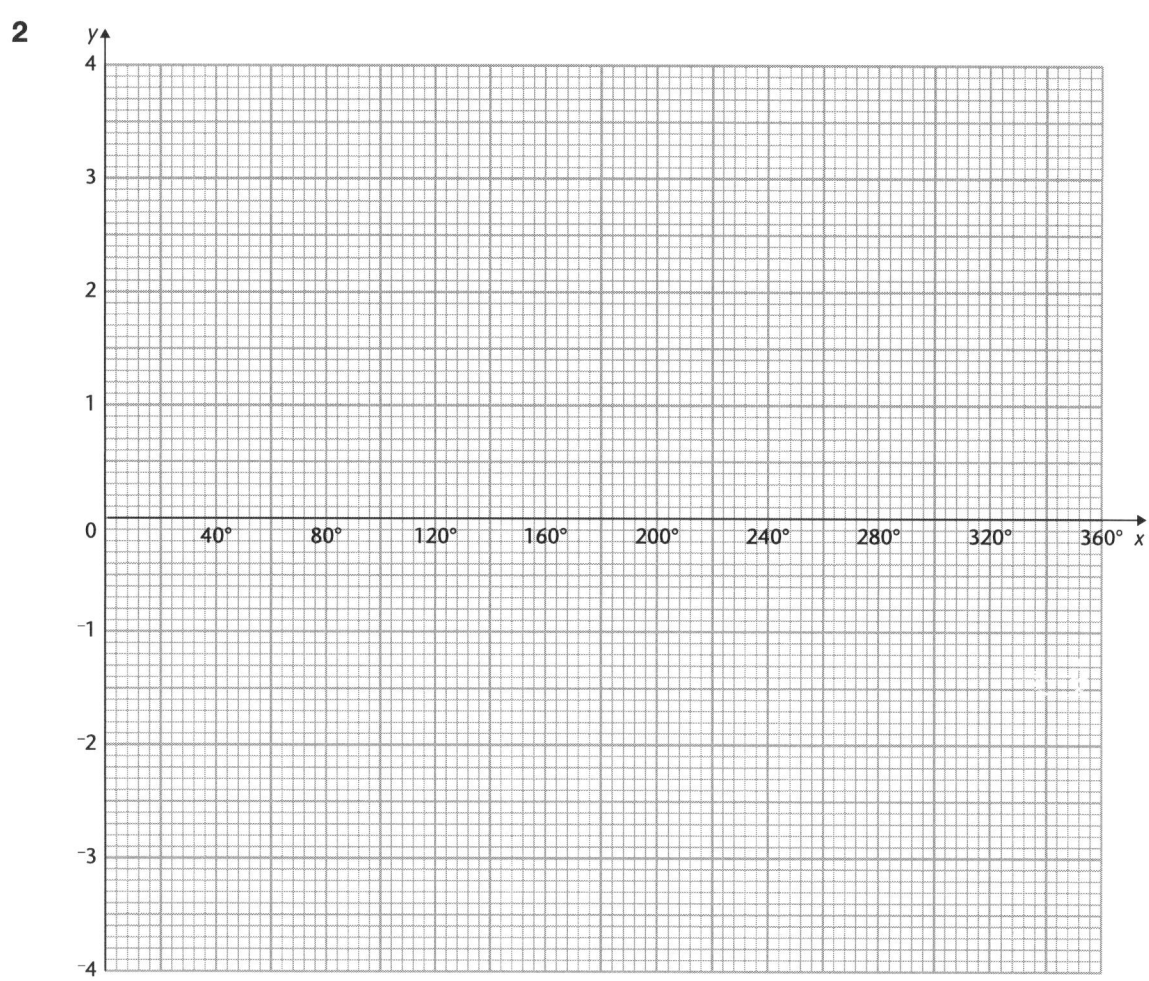

Graduated Assessment for OCR GCSE Mathematics © Hodder Murray 2007

STAGE
10

Exercise 13.2 (page 307)

1

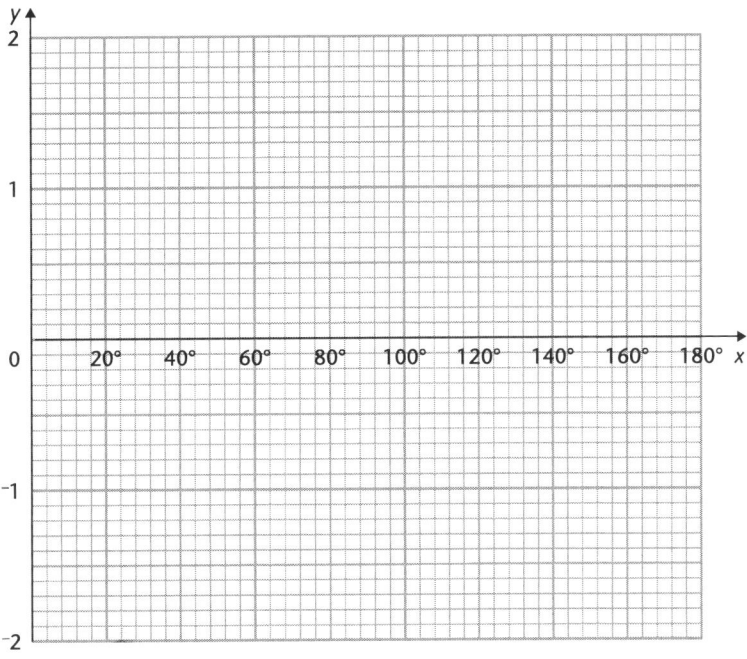

2

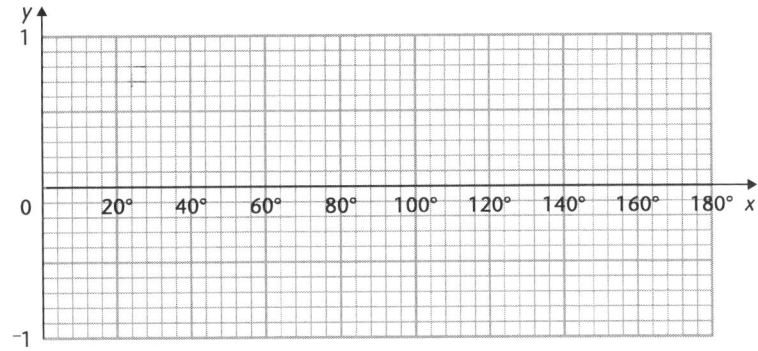

STAGE
10

Speed-up sheets

SPEED-UP SHEET 13.2 continued

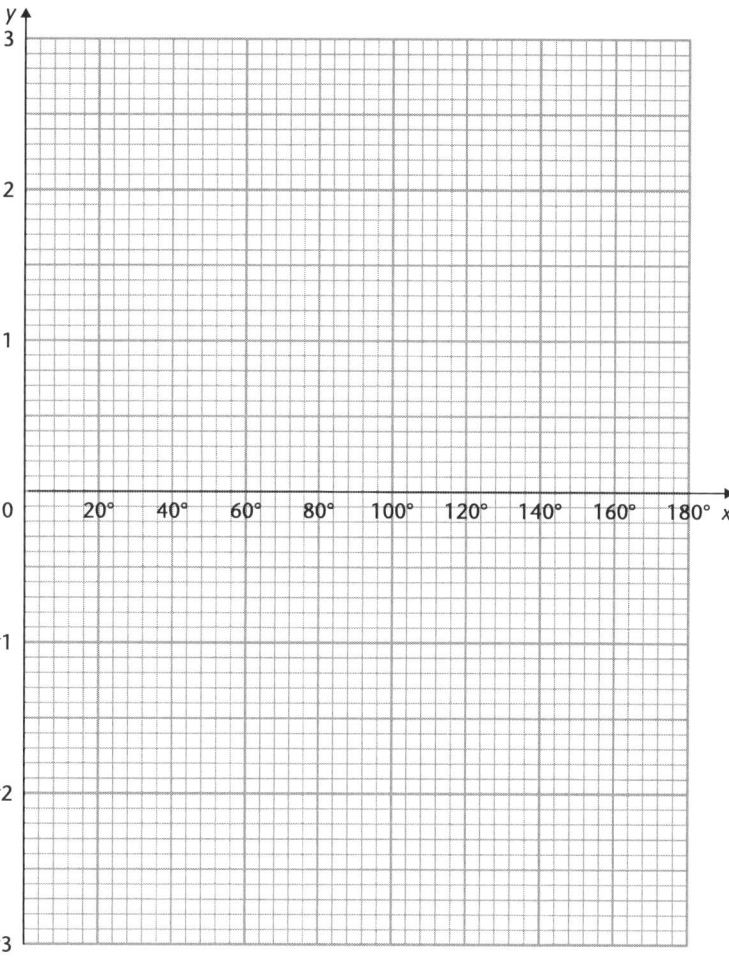

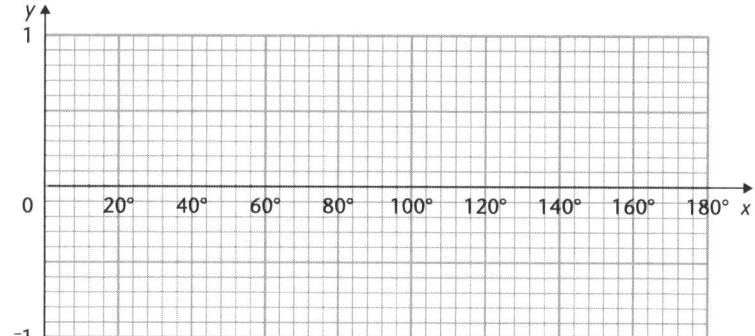

STAGE
10

Graduated Assessment for OCR GCSE Mathematics © Hodder Murray 2007

Stage 10 Answers

1 Using graphs to solve equations

Exercise 1.1 (page 157)

1 a) $y = x^2 - 5x + 5$

x	-2	-1	0	1	2	3	4	5
y	19	11	5	1	-1	-1	1	5

b) $2x + y = 9$

x	-2	0	5
y	13	9	-1

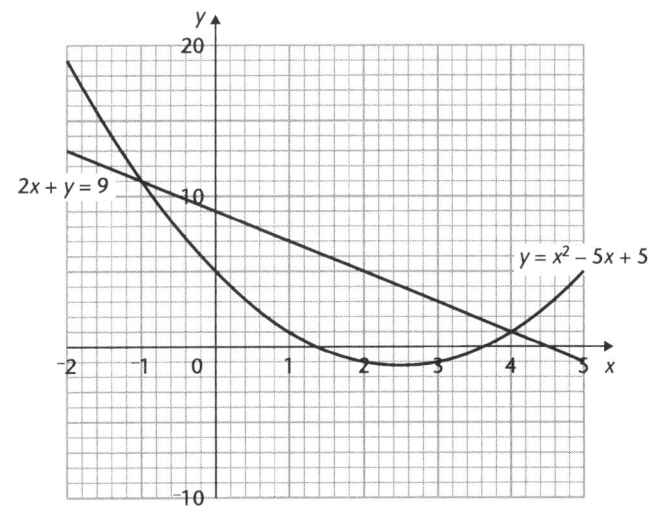

c) Cross at (-1, 11), (4, 1)

2 a) $y = x^2 - 3x - 1$

x	-4	-3	-2	-1	0	1	2	3
y	27	17	9	3	-1	-3	-3	-1

b) $4x + y = 5$

x	-4	0	3
y	21	5	-7

STAGE
10

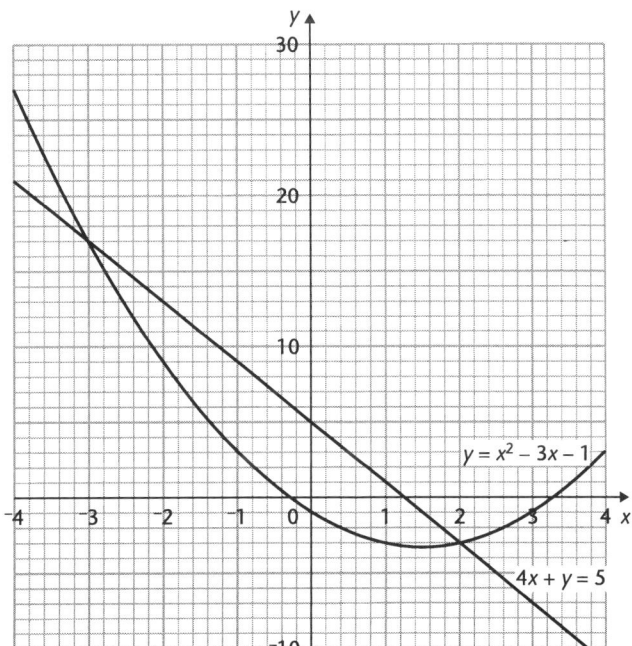

c) Cross at (-3, 17), (2, -3)

3 a) $y = x^2 + 3$

x	-2	-1	0	1	2	3	4	5
y	7	4	3	4	7	12	19	28

b) $y = 3x + 7$

x	-2	0	5
y	1	7	22

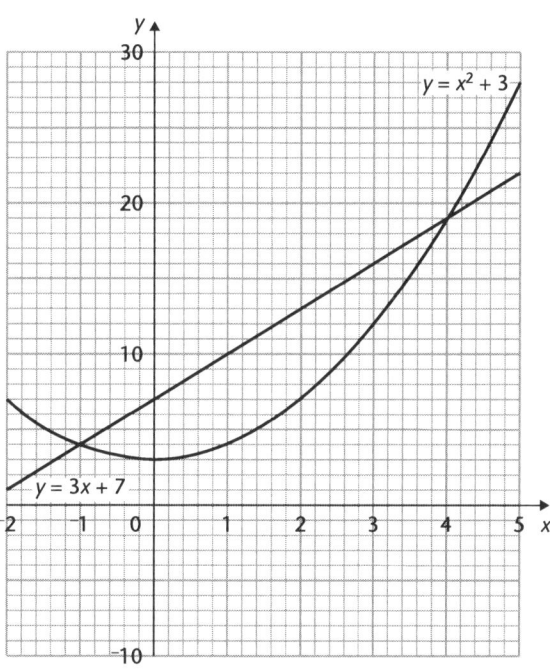

c) Cross at $(-1, 4)$, $(4, 19)$

4 a) $y = x^2 - 5x + 3$

x	-2	-1	0	1	2	3	4
y	17	9	3	-1	-3	-3	-1

b) $7x + 2y = 11$

x	-2	0	4
y	12·5	5·5	-8·5

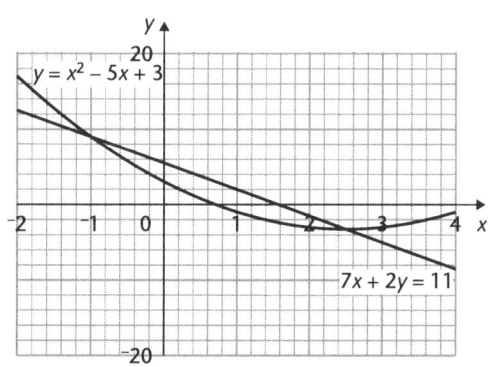

c) Cross at $(-1, 9)$, $(2·5, -3·2)$

5

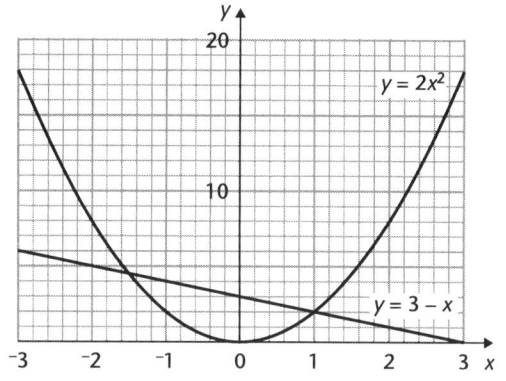

$x = 1$, $y = 2$ and $x = -1·5$, $y = 4·5$

6

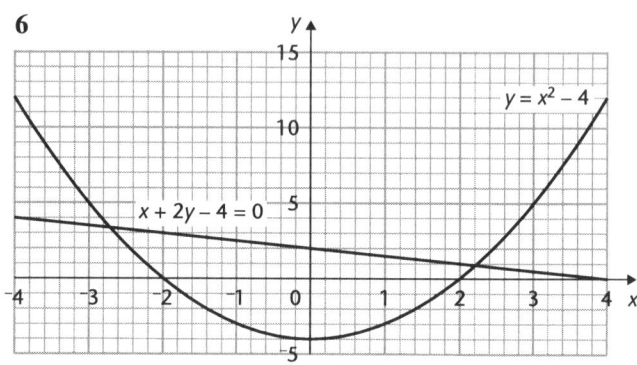

$x = 2·2$, $y = 0·9$ and $x = -2·7$, $y = 3·4$

Exercise 1.2 (page 162)

1 a)

x	-3	-2	-1	0	1	2	3
x^2	9	4	1	0	1	4	9
-5	-5	-5	-5	-5	-5	-5	-5
$y = x^2 - 5$	4	-1	-4	-5	-4	-1	4

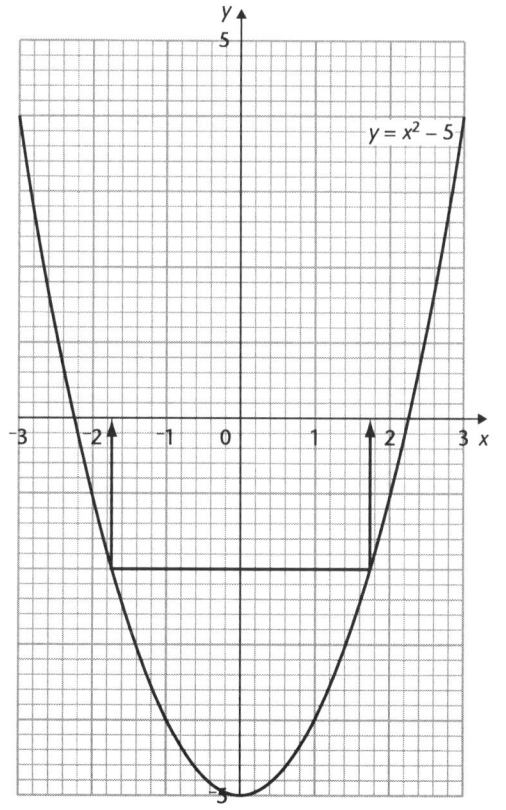

b) (i) $x = ^-2\cdot2$ or $2\cdot2$

(ii) Intersection with $y = ^-2$; $x = ^-1\cdot7$ or $1\cdot7$

2 a)

x	$^-2$	$^-1$	0	1	2	3	4
x^2	4	1	0	1	4	9	16
$2x^2$	8	2	0	2	8	18	32
$-3x$	6	3	0	$^-3$	$^-6$	$^-9$	$^-12$
$y = 2x^2 - 3x$	14	5	0	$^-1$	2	9	20

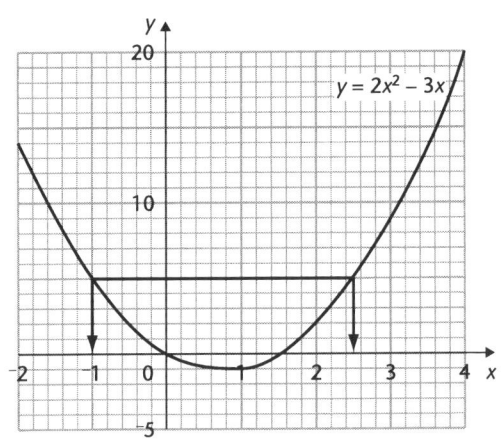

b) Solution of $2x^2 - 3x - 5 = 0$ is when $y = 5$; $x = ^-1$ or $2\cdot5$.

3 a)

x	$^-2$	$^-1$	0	1	2	3	4
x^2	4	1	0	1	4	9	16
5	5	5	5	5	5	5	5
$y = x^2 + 5$	9	6	5	6	9	14	21

For $y = 3x + 7$, three points are $(^-2, 1)$, $(0, 7)$, $(4, 19)$.

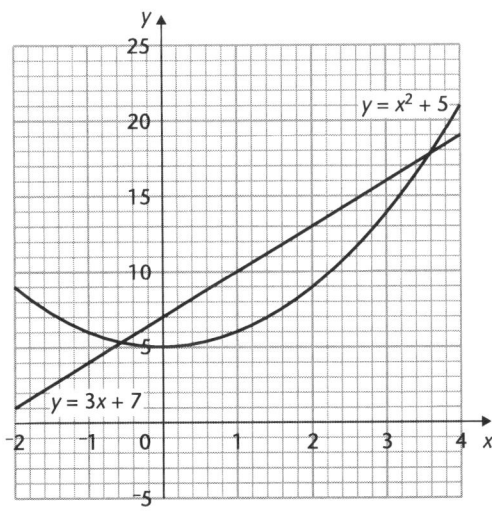

b) At intersection $x^2 + 5 = 3x + 7$ or $x^2 - 3x - 2 = 0$.

c) Solution is $x = ^-0\cdot6$ or $3\cdot6$

4 a)

x	$^-3$	$^-2$	$^-1$	0	1	2	3
x^2	9	4	1	0	1	4	9
$2x^2$	18	8	2	0	2	8	18
-10	$^-10$	$^-10$	$^-10$	$^-10$	$^-10$	$^-10$	$^-10$
$y = 2x^2 - 10$	8	$^-2$	$^-8$	$^-10$	$^-8$	$^-2$	8

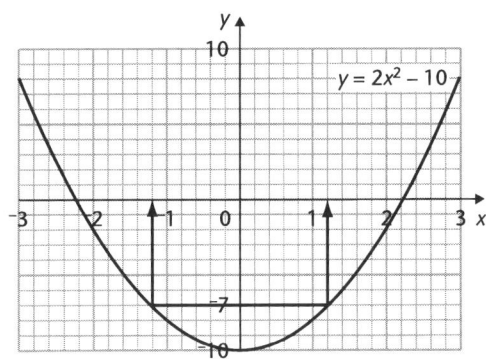

b) (i) When $y = 0$, $x = ^-2\cdot2$ or $2\cdot2$.

(ii) $2x^2 - 3 = 0$ is the same as $2x^2 - 10 + 7 = 0$ or $2x^2 - 10 = ^-7$; where $y = ^-7$, $x = ^-1\cdot2$ or $1\cdot2$

5 a)

x	$^-2$	$^-1$	0	1	2	3	4
x^2	4	1	0	1	4	9	16
2	2	2	2	2	2	2	2
$y = x^2 + 2$	6	3	2	3	6	11	18

$y = 2x + 7$ points are $(^-2, 3)$, $(0, 7)$, $(4, 15)$

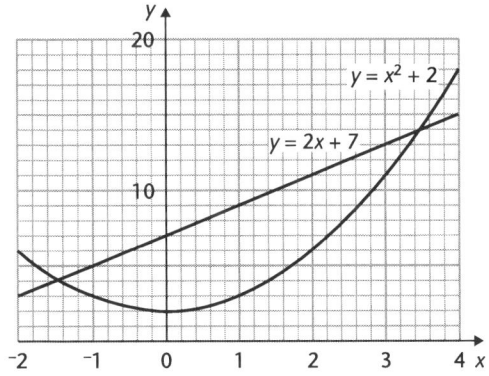

b) At intersection $x^2 + 2 = 2x + 7$, or $x^2 - 2x - 5 = 0$.

c) Solution is $x = ^-1\cdot5$ or $3\cdot5$

6 a)

x	$^-1$	0	1	2	3	4	5	6
x^2	1	0	1	4	9	16	25	36
$-5x$	5	0	$^-5$	$^-10$	$^-15$	$^-20$	$^-25$	$^-30$
$y = x^2 - 5x$	6	0	$^-4$	$^-6$	$^-6$	$^-4$	0	6

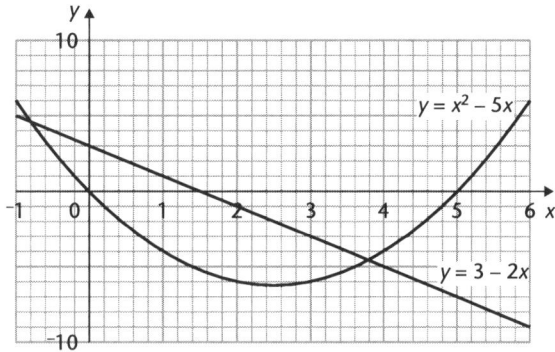

b) (i) $x^2 - 3x = 0$ is the same as
$x^2 - 5x + 2x - 3 = 0$ or $x^2 - 5x = 3 - 2x$,
so other graph is $y = 3 - 2x$.
(ii) Solution is $x = ^-0{\cdot}8$ or $3{\cdot}8$

7 a)

x	$^-3$	$^-2$	$^-1$	0	1	2	3
x^3	$^-27$	$^-8$	$^-1$	0	1	8	27
$-3x$	9	6	3	0	$^-3$	$^-6$	$^-9$
$y = x^3 - 3x$	$^-18$	$^-2$	2	0	$^-2$	2	18

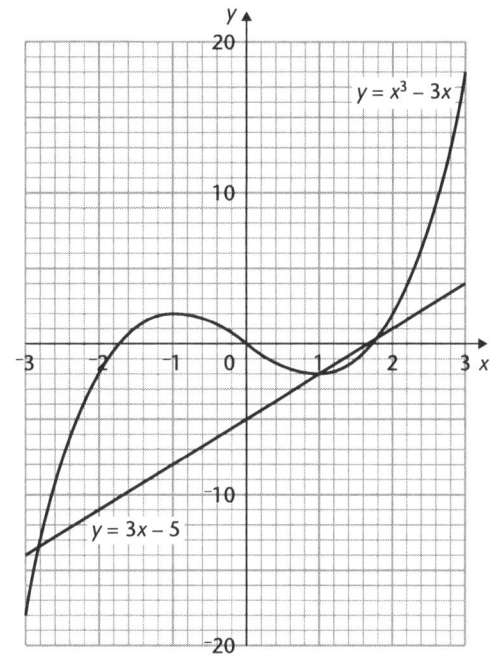

b) (i) $x^3 - 6x + 5 = 0$ is the same as
$x^3 - 3x - 3x + 5 = 0$ or
$x^3 - 3x = 3x - 5$, so other graph is
$y = 3x - 5$.
(ii) Solution is $x = ^-2{\cdot}8$ or 1 or $1{\cdot}8$

8 a)

x	$^-3$	$^-2$	$^-1$	0	1	2	3
x^3	$^-27$	$^-8$	$^-1$	0	1	8	27
$-5x$	15	10	5	0	$^-5$	$^-10$	$^-15$
$y = x^3 - 5x$	$^-12$	2	4	0	$^-4$	$^-2$	12

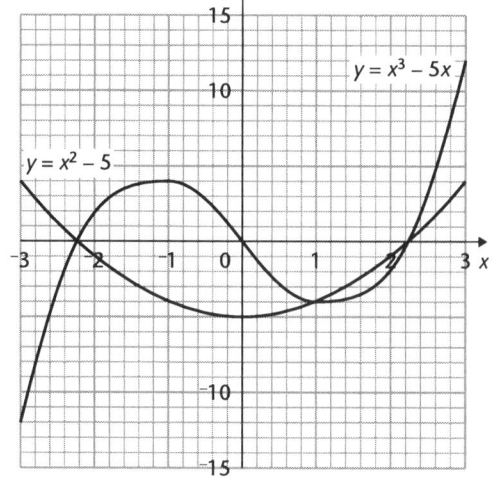

b) (i) $x^3 - x^2 - 5x + 5 = 0$ is the same as
$x^3 - 5x = x^2 - 5$, so other graph is
$y = x^2 - 5$.

x	$^-3$	$^-2$	$^-1$	0	1	2	3
x^2	9	4	1	0	1	4	9
-5	$^-5$	$^-5$	$^-5$	$^-5$	$^-5$	$^-5$	$^-5$
$y = x^2 - 5$	4	$^-1$	$^-4$	$^-5$	$^-4$	$^-1$	4

(ii) Solution is $x = ^-2{\cdot}2$, 1, $2{\cdot}2$

STAGE 10

9 a)

x	$^-2$	$^-1$	0	1	2	3	4
x^2	4	1	0	1	4	9	16
$-2x$	4	2	0	$^-2$	$^-4$	$^-6$	$^-8$
-4	$^-4$	$^-4$	$^-4$	$^-4$	$^-4$	$^-4$	$^-4$
$y = x^2 - 2x - 4$	4	$^-1$	$^-4$	$^-5$	$^-4$	$^-1$	4

10 a)

x	$^-6$	$^-5$	$^-4$	$^-3$	$^-2$	$^-1$	0	1
x^2	36	25	16	9	4	1	0	1
$5x$	$^-30$	$^-25$	$^-20$	$^-15$	$^-10$	$^-5$	0	5
4	4	4	4	4	4	4	4	4
$y = x^2 + 5x + 4$	10	4	0	$^-2$	$^-2$	0	4	10

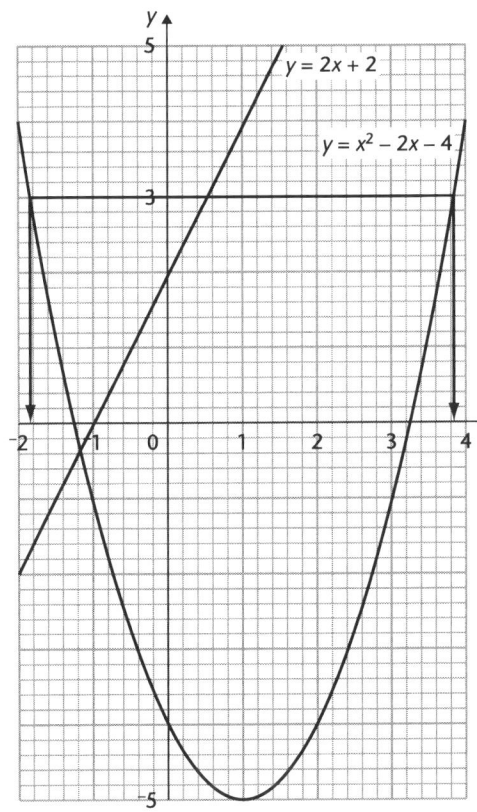

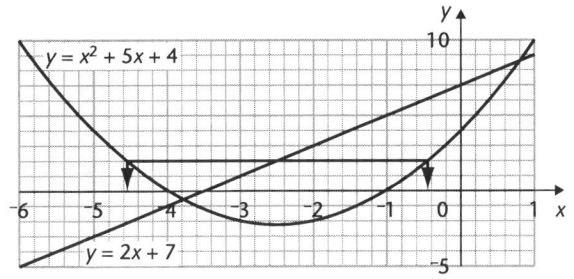

b) (i) $x = ^-4$ or $^-1$

(ii) $x^2 + 3x - 3 = 0$ is the same as
$x^2 + 5x + 4 - 2x - 7 = 0$, or
$x^2 + 5x + 4 = 2x + 7$.
Draw $y = 2x + 7$,
solution is $x = ^-3\cdot8$ or $0\cdot8$.

(iii) $x^2 + 5x + 2 < 0$ is the same as
$x^2 + 5x + 4 - 2 < 0$ or $x^2 + 5x + 4 < 2$,
where $y < 2$.
Solution is $^-4\cdot6 < x < ^-0\cdot4$

11 Find where the curve crosses
 a) $y = 0$
 b) $y = 2$

12 Find where the curve crosses $y = 5$.

13 Find where the curve crosses
 a) $y = 0$
 b) $y = ^-4$

14 Find where the curve crosses $y = 3$.

15 $x^2 - 4x + 3 = 0$

16 $y = x^2 - 3$

17 $x^2 - x + 3 = 0$

18 $y = ^-x^2 + 4x - 3$

19 $y = 3 - x$

20 $y = 4x + 4$

b) (i) $x^2 - 2x - 7 = 0$ is the same as
$x^2 - 2x - 4 - 3 = 0$ or $x^2 - 2x - 4 = 3$;
where $y = 3$, $x = ^-1\cdot8$ or $3\cdot8$.

(ii) $x^2 - 4x - 6 = 0$ is the same as
$x^2 - 2x - 4 - 2x - 2 = 0$ or
$x^2 - 2x - 4 = 2x + 2$, so other graph is
$y = 2x + 2$, so $x = ^-1\cdot2$.
The curve and the line will also cross
at $(5\cdot2, 12\cdot4)$ but this is not on the
graph drawn.

(iii) $x^2 - 2x - 4 > 0$ above x-axis;
$x < ^-1\cdot2$ or $x > 3\cdot2$

Exercise 1.3 (page 166)

1

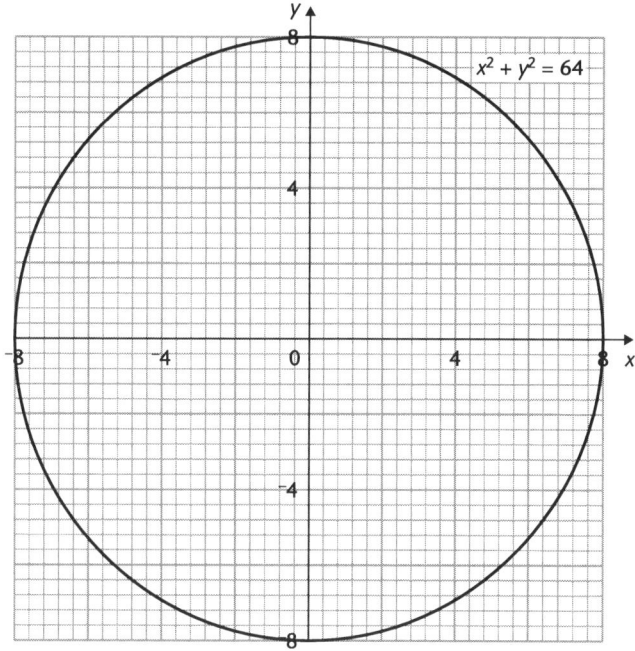

2

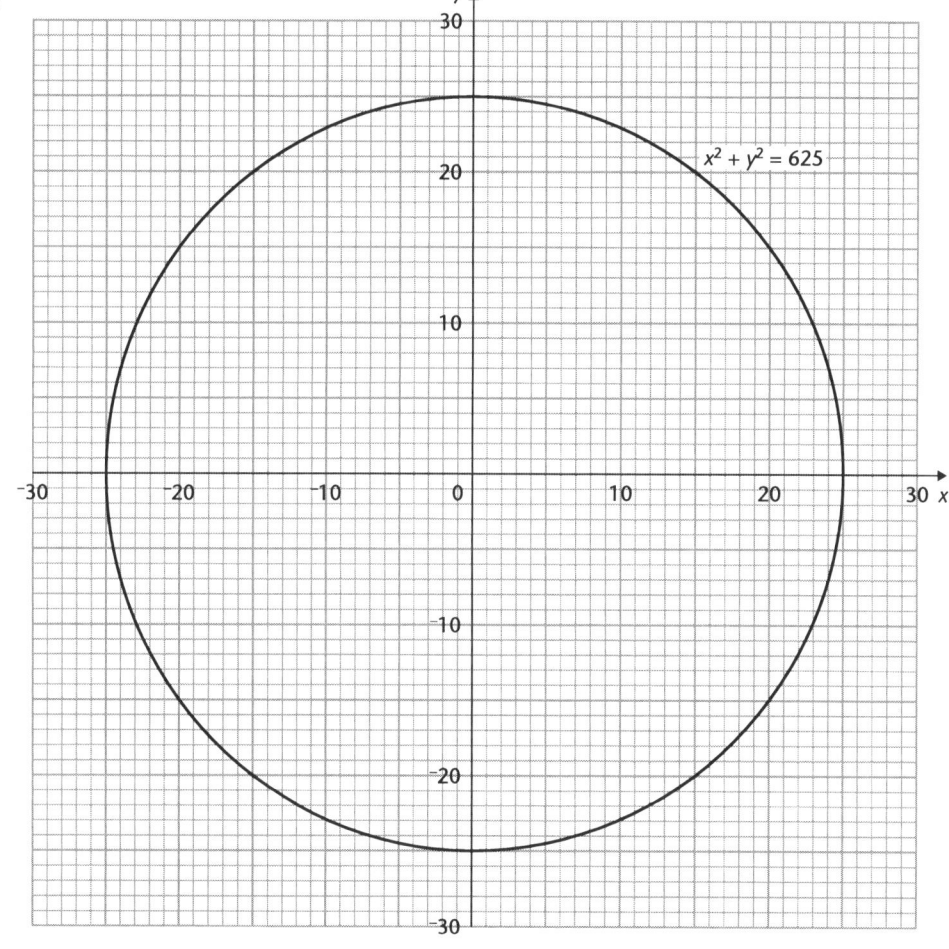

 Graduated Assessment for OCR GCSE Mathematics © Hodder Murray 2007

3 a)

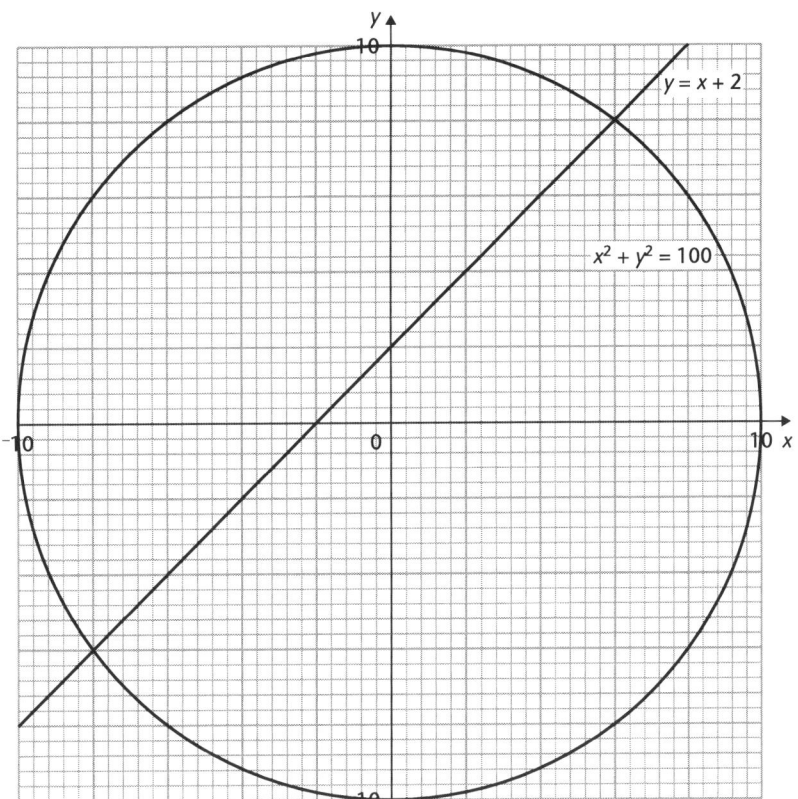

$y = x + 2$

$x^2 + y^2 = 100$

b) Cross at $(^-8, ^-6)$ and $(6, 8)$

4 a)

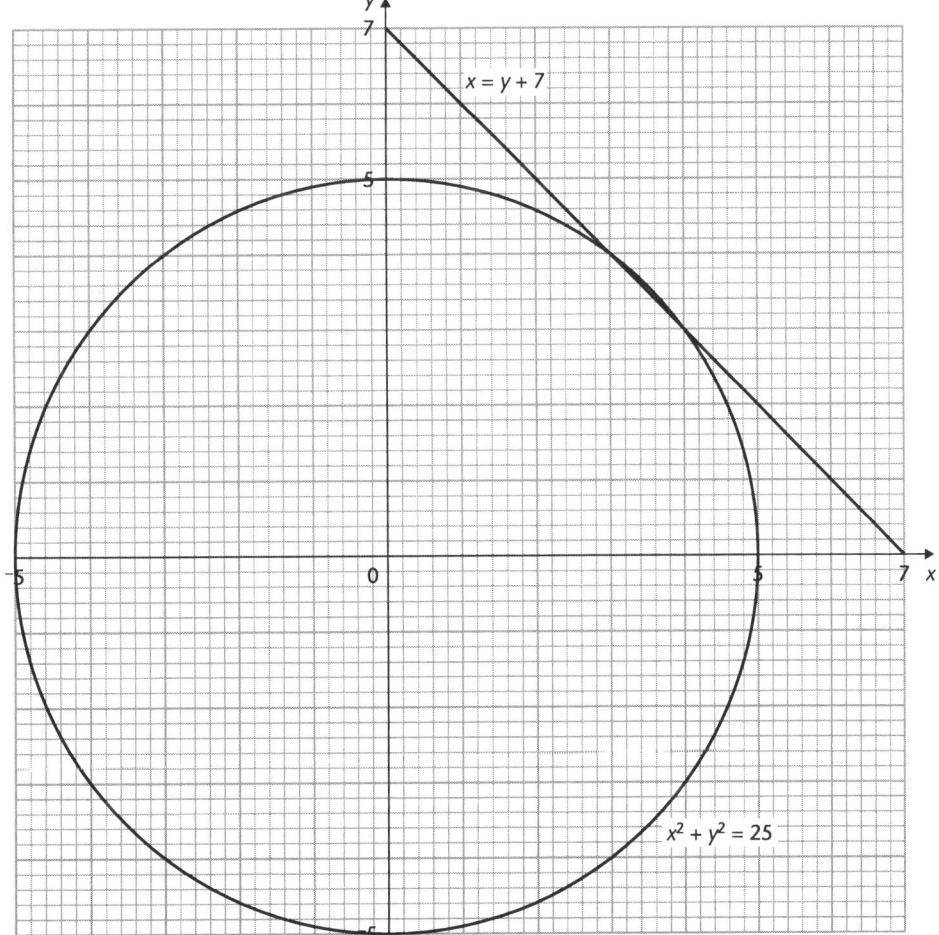

$x = y + 7$

$x^2 + y^2 = 25$

b) Cross at $(3, 4)$ and $(4, 3)$

STAGE
10

5 a)

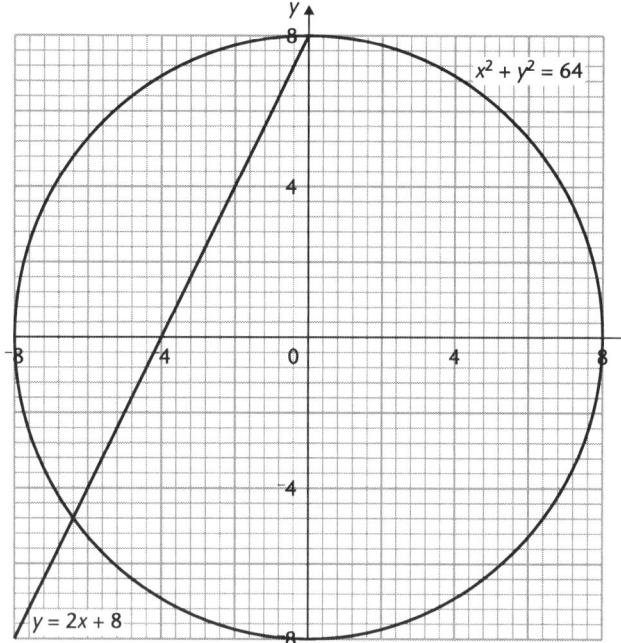

b) $x = {}^-6.4$, $y = {}^-4.8$ or $x = 0$, $y = 8$

6 a)

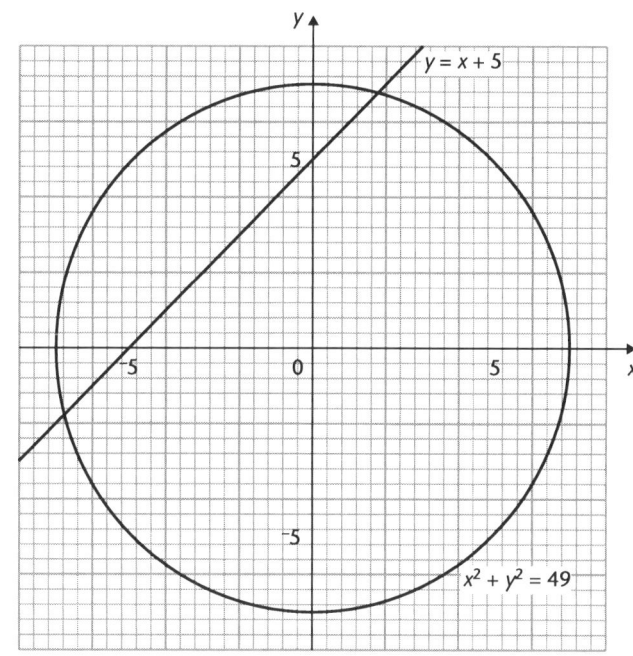

b) $x = {}^-6.8$, $y = {}^-1.8$ or $x = 1.8$, $y = 6.8$

2 Growth and decay

Exercise 2.1 (page 174)

1 $n = 2 \times 3^t$

2 $m = 100 \times 0.5^t$

3 $y = 2^x$

x	-2	-1	0	1	2	3	4	5
y	0·25	0·5	1	2	4	8	16	32

Graph shown half-size.

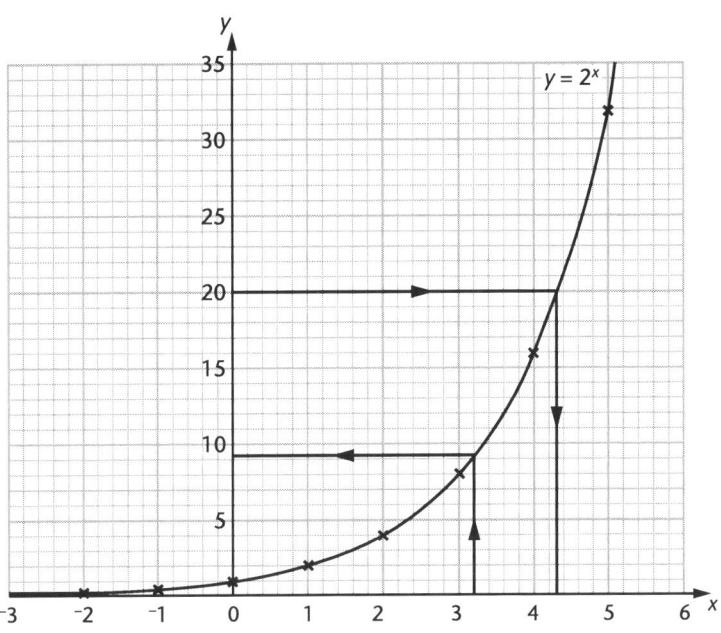

a) $y = 9·2$
b) $x = 4·3$

4 $y = 1·5^x$

x	-3	-2	-1	0	1	2	3	4	5
y	0·30	0·44	0·67	1	1·5	2·25	3·38	5·06	7·59

Graph shown half-size.

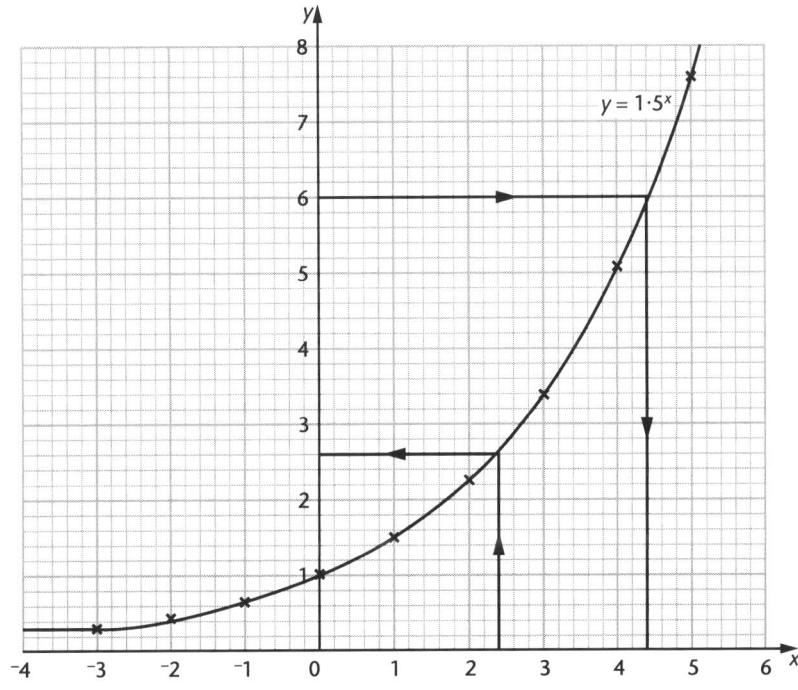

a) $y = 2·6$
b) $x = 4·4$

5 $y = 3^{-x}$

x	0	0·5	1	1·5	2	2·5	3	3·5	4
y	1	0·577	0·333	0·192	0·111	0·064	0·037	0·021	0·012

Graph shown half-size.

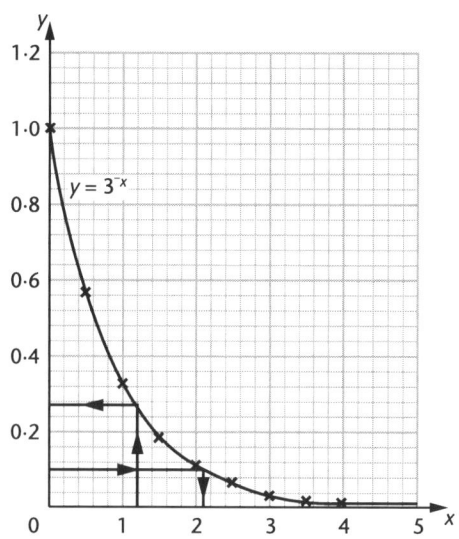

a) $y = 0·27$
b) $x = 2·1$

6 $y = 4^{-x}$

x	-2·5	-2	-1·5	-1	-0·5	0	0·5	1
y	32	16	8	4	2	1	0·5	0·25

Graph shown half-size.

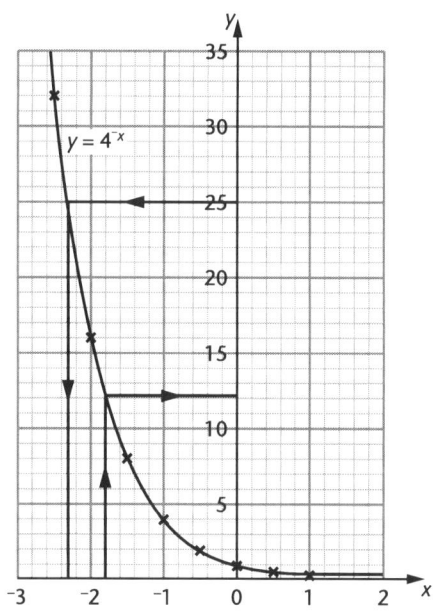

a) $y = 12·1$
b) $x = -2·3$

Exercise 2.2 (page 177)

1 a) 702·46, 786·76, 881·2
   b) 3·6
2 a) 1, 3, 9, 27, 81, 243
   b) 531 441
   c) 7·9
3 a)

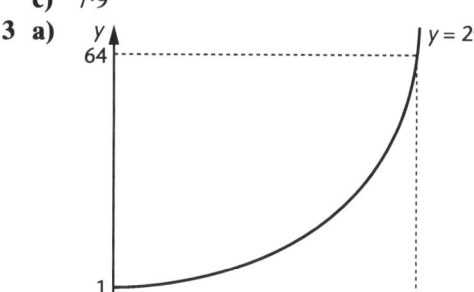

   b) 7·6
4 a) 100, 200, 400, 800, 1600, 3200, 6400, 12 800
   b) $8·4 \times 10^8$
   c) 1·6 hours
5 a) 25 **b)** 9510 to 3 s.f.
   c) 35 minutes
6 a) £103·32 **b)** $£85 \times 1·05^x$
   c) 14(·2) or 15 years
7 3·4 years
8 6·6 years
9 a) £9030·56 **b)** 30 years
10 a) $£16 000 \times 0·84^n$ **b)** 6·7 years
11 $a = 30, b = 2$
12 a) £1312·20 **b)** 6·6 years
13 a)

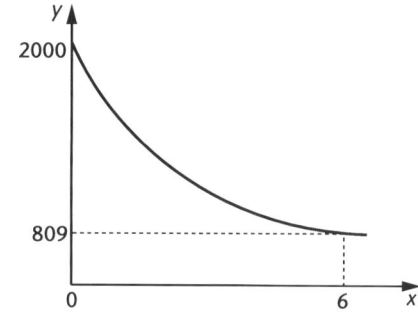

   b) 6·5 hours
14 15·7 years
15 a) $£2000 \times 1·06^n$ **b)** 19 years
16 a) 100 g **b)** 2·32
17 a) $v = 9000 \times 0·88^t$
   b) **(i)** £6133·25 **(ii)** £3236·71
   c) 5 years
18 a) 1 000 000
   b) **(i)** 31 250 **(ii)** 244
   c) 20 hours
19 $a = 5, b = 1·6$
20 a) $x = 4·32$ **b)** $x = 0·63$
   c) $x = 4·42$ **d)** $x = 1·66$
   e) $x = -1·96$

3 Rational and irrational numbers

Exercise 3.1 (page 185)

1. a) R, fraction
   b) R, terminating decimal
   c) I, $\sqrt{2}$ is irrational
   d) R, $\sqrt{169} = 13$
   e) I, π is irrational
   f) R, terminating decimal
   g) R, recurring decimal
   h) I, $\sqrt{3}$ is irrational
   i) I, $\sqrt{2}$ is irrational
   j) R, fraction
   k) R, terminating decimal
   l) R, recurring decimal
   m) R, $\sqrt{324} = 18$
   n) I, cannot be written as a fraction
   o) I, π is irrational
   p) R, terminating decimal
   q) R, π cancels
   r) I, $\sqrt{3}$ and $\sqrt{5}$ are irrational
   s) I, $\sqrt{2}$ is irrational
   t) R, fraction

2. a) T, 5 is a factor of 10
   b) R, 17 is prime
   c) T, $125 = 5^3$ and 5 is a factor of 10
   d) R, $18 = 2 \times 3^2$ and 3 is not a factor of 10
   e) T, $8 = 2^3$ and 2 is a factor of 10
   f) R, $15 = 3 \times 5$ and 3 is not a factor of 10
   g) T, $20 = 2^2 \times 5$ and 2 and 5 are factors of 10
   h) R, $35 = 5 \times 7$ and 7 is not a factor of 10
   i) T, $\frac{9}{120} = \frac{3}{40}$ and 40's prime factors are all factors of 10
   j) R, $12 = 2^2 \times 3$ and 3 is not a factor of 10

3. a) $\frac{3}{25}$ b) $\frac{41}{200}$
   c) $\frac{3}{8}$ d) $\frac{5}{16}$

4. a) $0.6\dot{3}$ b) $0.\dot{4}2857\dot{1}$
   c) $0.04\dot{2}857\dot{1}$ d) $0.2\dot{5}$
   e) $0.41\dot{6}\dot{2}$ f) $0.2\dot{1}$
   g) 0.384615 h) $0.003846\dot{1}\dot{5}$
   i) $0.47\dot{2}$ j) $0.43\dot{7}\dot{2}$

5. a) $\frac{2}{9}$ b) $\frac{7}{9}$
   c) $\frac{16}{33}$ d) $\frac{7}{30}$
   e) $\frac{44}{333}$ f) $\frac{43}{99}$
   g) $\frac{134}{333}$ h) $\frac{13}{55}$
   i) $\frac{35}{99}$ j) $\frac{6}{11}$
   k) $\frac{4}{33}$ l) $\frac{17}{99}$
   m) $\frac{1234}{9999}$ n) $2\frac{2}{11}$

Exercise 3.2 (page 189)

1. a) $2\sqrt{3}$, I   b) $10\sqrt{10}$, I
   c) $3\sqrt{5}$, I   d) $10\sqrt{3}$, I
   e) $5\sqrt{3}$, I   f) 4, R
   g) $6\sqrt{10}$, I   h) 2, R
   i) $20\sqrt{10}$, I   j) $15\sqrt{5}$, I
   k) $2\sqrt{10}$, I   l) $3\sqrt{6}$, I
   m) $7\sqrt{2}$, I   n) $20\sqrt{2}$, I
   o) $11\sqrt{3}$, I   p) 9, R
   q) 100, R   r) 3, R
   s) $30\sqrt{2}$, I   t) 10, R

2. a) 8   b) $2\sqrt{3}$
   c) 13

3. a) 8   b) $2 + 2\sqrt{7}$
   c) $8 - 2\sqrt{7}$

4. a) $7 - 2\sqrt{5}$ b) $^-1 + 4\sqrt{5}$
   c) $14 + 6\sqrt{5}$

5. a) $13 - \sqrt{11}$ b) $^-5 + 3\sqrt{11}$
   c) $27 + 8\sqrt{11}$

6. a) $5\sqrt{3} + 6$ b) $37 + 20\sqrt{3}$
   c) $34 - 24\sqrt{2}$

7. a) $6\sqrt{5} - 10$ b) $3\sqrt{3} - 15$
   c) $56 - 24\sqrt{5}$ d) $84 - 30\sqrt{3}$

8. $60 + 43\sqrt{2}$

9. 37

10. a) $\frac{\sqrt{2}}{2}$ b) $\frac{2\sqrt{5}}{5}$
    c) $\frac{5\sqrt{7}}{7}$ d) $\frac{11\sqrt{2}}{6}$
    e) $\frac{9\sqrt{5}}{10}$ f) $\frac{\sqrt{7}}{7}$
    g) $\frac{3\sqrt{2}}{2}$ h) $\frac{5\sqrt{11}}{11}$
    i) $\frac{7\sqrt{2}}{10}$ j) $\frac{9\sqrt{2}}{8}$

STAGE
10

11 a) $\dfrac{3\sqrt{2}}{2}$ b) $\dfrac{\sqrt{3}}{5}$

  c) $\dfrac{4\sqrt{3}}{5}$ d) $\dfrac{2\sqrt{6}}{3}$

  e) $2\sqrt{5}$ f) $\dfrac{3\sqrt{2}}{2}$

  g) $\dfrac{5\sqrt{2}}{2}$ h) $\dfrac{6\sqrt{5}}{5}$

  i) $5\sqrt{2}$ j) $\dfrac{\sqrt{10}}{5}$

  k) $\dfrac{2\sqrt{10}}{15}$ l) $\dfrac{7\sqrt{2}}{10}$

  m) $\dfrac{\sqrt{6}}{3}$ n) $\dfrac{12\sqrt{10}}{35}$

12 a) $3 + 3\sqrt{2}$ b) $\dfrac{1 + 3\sqrt{5}}{2}$

  c) $\dfrac{4\sqrt{3} + 6}{2}$ d) $\dfrac{5\sqrt{6} + 6\sqrt{2}}{6}$

13 10
14 7π
15 $\sqrt{10}$
16 $12 + \dfrac{3\pi}{2}$ or $12 + 1\cdot5\pi$

4 Trigonometry in non-right-angled triangles

Exercise 4.1 (page 196)

1 $c = 5\cdot39\,\text{cm}$, $A = 46°$, $a = 5\cdot22\,\text{cm}$
2 $p = 11\cdot6\,\text{cm}$, $R = 26°$, $r = 5\cdot50\,\text{cm}$
3 $g = 14\cdot6\,\text{cm}$, $E = 55°$, $e = 15\cdot2\,\text{cm}$
4 $b = 271\,\text{m}$, $C = 68\cdot5°$, $c = 260\,\text{m}$
5 $B = 66°$, $C = 72°$, $c = 7\cdot39\,\text{cm}$
6 $M = 71\cdot4°$, $N = 28\cdot6°$, $n = 6\cdot46\,\text{cm}$
7 $E = 34\cdot2°$, $D = 120°$, $d = 9\cdot11\,\text{cm}$
8 $A = 42\cdot9°$, $B = 66\cdot1°$, $b = 96\cdot7\,\text{m}$
9 $P = 32\cdot2°$, $R = 78\cdot4°$, $r = 7\cdot53\,\text{cm}$
10 $Y = 35\cdot5°$, $Z = 48\cdot5°$, $z = 9\cdot04\,\text{cm}$
11 $T = 68\cdot8°$, $S = 80°$, $s = 9\cdot50\,\text{m}$
12 $b = 12\cdot5\,\text{mm}$, $C = 47°$, $c = 9\cdot28\,\text{mm}$
13 $y = 7\cdot10\,\text{cm}$, $Z = 45°$, $z = 7\cdot81\,\text{cm}$
14 $s = 1\cdot13\,\text{m}$, $T = 59°$, $t = 2\cdot70\,\text{m}$
15 $28\cdot2\,\text{cm}$
16 $15\cdot7\,\text{cm}$
17 $B = 94\cdot3°$
18 a) $AT = 85\cdot7\,\text{m}, BT = 60\cdot5\,\text{m}$
  b) $38\cdot9\,\text{m}$
19 a) $AB = 25\cdot7\,\text{m}, BC = 42\cdot7\,\text{m}$
  b) $23\cdot9\,\text{m}$
20 $380\,\text{m}$
21 $AC = 43\cdot9\,\text{km}$, $BC = 25\cdot3\,\text{km}$
22 $6\cdot0\,\text{m}$
23 a) $54°$ b) $46°$

Exercise 4.2 (page 201)

1 $14\cdot2\,\text{cm}$
2 $3\cdot91\,\text{cm}$
3 $13\cdot7\,\text{m}$
4 $21\cdot6\,\text{cm}$
5 $48\cdot5°$
6 $50\cdot7°$
7 $110\cdot7°$
8 $8\cdot89\,\text{cm}$
9 $18\cdot7\,\text{cm}$
10 $14\cdot2\,\text{cm}$
11 $5\cdot37\,\text{cm}$
12 $52\cdot0°$
13 $39\cdot5°$
14 $49\cdot3°$
15 $A = 45\cdot9°$ (opposite the shortest side)
16 $A = 45\cdot7°, B = 62\cdot5°, C = 71\cdot7°$
17 $4\cdot79\,\text{km}$
18 $9\cdot6\,\text{km}$
19 $x = 11\cdot3\,\text{m}, y = 19\cdot5°$
20 $4\cdot85\,\text{cm}, 6\cdot40\,\text{cm}$
21 a) $45\cdot8\,\text{m}$ b) $110\cdot1°$
22 a) (i) $11\cdot7\,\text{m}$ (ii) $10\cdot2\,\text{m}$ (iii) $10\cdot8\,\text{m}$
  b) (i) $58\cdot6°$ (ii) $67\cdot6°$

Exercise 4.3 (page 205)

1 a) $8\cdot94\,\text{cm}^2$ b) $19\cdot7\,\text{cm}^2$
  c) $20\cdot5\,\text{cm}^2$ d) $5\cdot20\,\text{m}^2$
  e) $34\cdot0\,\text{cm}^2$ f) $12\cdot1\,\text{m}^2$
2 $15\,\text{cm}^2$
3 $73\cdot2°$
4 $44\cdot6\,\text{cm}$
5 $35\cdot7\,\text{cm}^2$

Revision exercise A1 (page 207)

1 a)

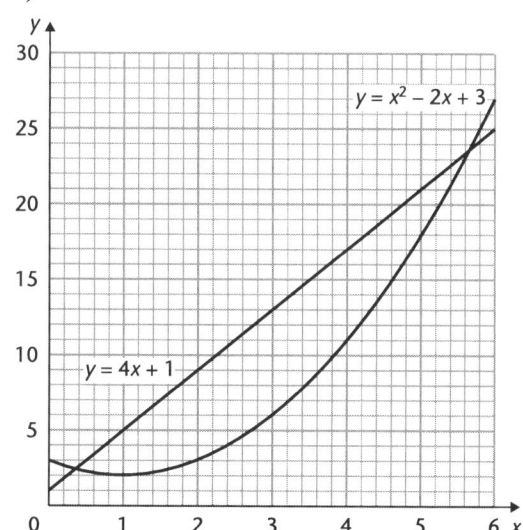

  b) $x = 0\cdot4, y = 2\cdot4$ or $x = 5\cdot6, y = 23\cdot6$

2 a)

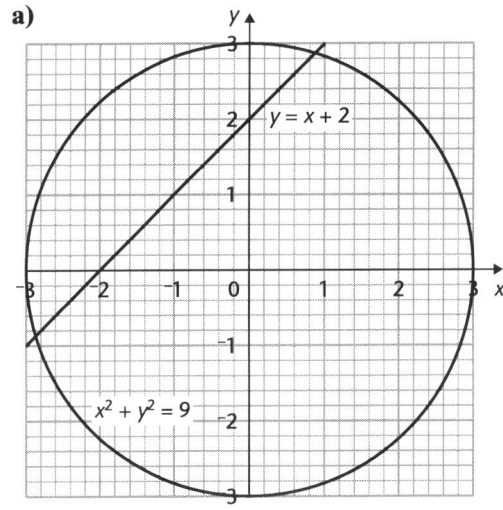

b) $x = {}^-2{\cdot}9$, $y = {}^-0{\cdot}9$ or $x = 0{\cdot}9$, $y = 2{\cdot}9$

3 $x = 0$, $y = 6$ or $x = {}^-6$, $y = 0$

4 a) $y = x^2 + 3x - 7$

x	$^-6$	$^-5$	$^-4$	$^-3$	$^-2$	$^-1$	0	1	2	3
x^2	36	25	16	9	4	1	0	1	4	9
$+ 3x$	$^-18$	$^-15$	$^-12$	$^-9$	$^-6$	$^-3$	0	3	6	9
$- 7$	$^-7$	$^-7$	$^-7$	$^-7$	$^-7$	$^-7$	$^-7$	$^-7$	$^-7$	$^-7$
$y = x^2 + 3x - 7$	11	3	$^-3$	$^-7$	$^-9$	$^-9$	$^-7$	$^-3$	3	11

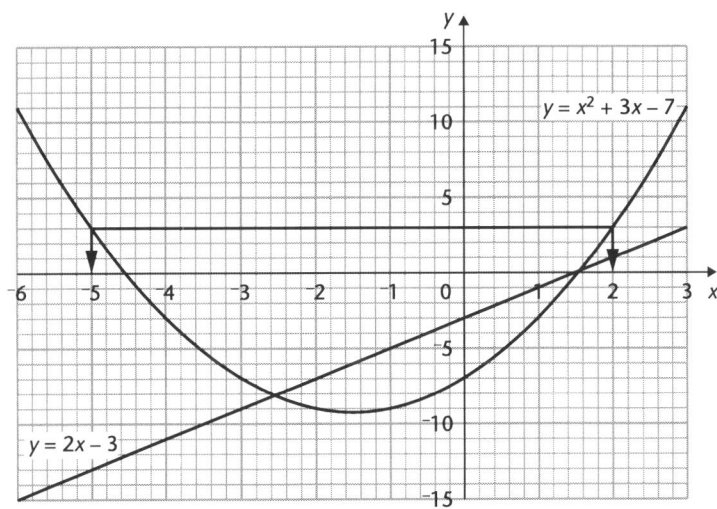

b) (i) $x^2 + 3x - 7 = 0$ when $y = 0$; solution is $x = {}^-4{\cdot}5$ or $1{\cdot}5$

(ii) $x^2 + 3x - 10 = 0$ is the same as $x^2 + 3x - 7 - 3 = 0$ or $x^2 + 3x - 7 = 3$.
When $y = 3$, $x = {}^-5$ or 2.

c) (i) $x^2 + x - 4 = 0$ is the same as $x^2 + 3x - 7 - 2x + 3 = 0$ or $x^2 + 3x - 7 = 2x - 3$.
Equation is $y = 2x - 3$.

(ii) Solution is $x = {}^-2{\cdot}6$ or $1{\cdot}6$

5 44.9 minutes

6 $y = 2^{-x}$

x	0	0·5	1	1·5	2	2·5	3	3·5	4
y	1	0·707	0·5	0·354	0·25	0·177	0·125	0·088	0·063

Graph shown half-size.

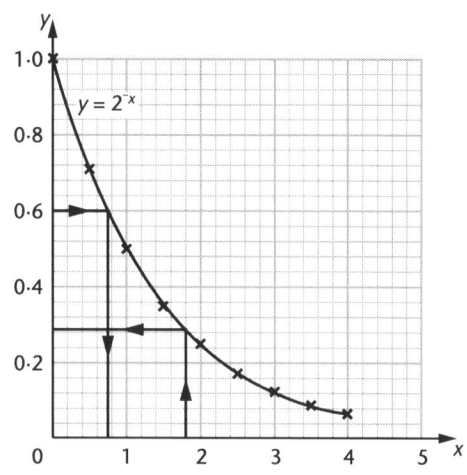

  a) $y = 0·29$
  b) $x = 0·74$
7 **a)** 100
  b) 249
  c) 12.6 days
8 $a = 10$, $b = 0·8$
9 **a)** R, ter141minating decimal
  b) R, recurring decimal
  c) I, π is irrational
  d) I, $\sqrt{3}$ is irrational
  e) R, terminating decimal
10 **a)** $0·4\dot{5}$
  b) $0·2\dot{1}\dot{2}$
  c) $0·0\dot{7}\dot{4}$
11 **a)** $\frac{6}{11}$
  b) $3\frac{73}{495}$
  c) $\frac{226}{1111}$

12 **a)** $4\sqrt{2}$
  b) $5\sqrt{6}$
  c) $8\sqrt{2}$
  d) 30
  e) $6\sqrt{5}$
  f) $2\sqrt{2}$
  g) $48\sqrt{6}$
13 **a)** $8 - 3\sqrt{7}$
  b) $^{-}2 + 5\sqrt{7}$
  c) $^{-}13 - 7\sqrt{7}$
14 **a)** $37 + 20\sqrt{3}$
  b) $37 - 20\sqrt{3}$
  c) 13
15 $65\sqrt{10} + 200$
16 **a)** $\dfrac{11\sqrt{2}}{2}$
  b) $\dfrac{5\sqrt{3}}{2}$
  c) $\dfrac{2\sqrt{3}}{3}$
17 **a)** 12·3°
  b) 132°
  c) 8·30 m
18 **a)** 13 cm
  b) 32·2°
  c) 52·0 cm²
19 **a)** 38·2°
  b) 60°
20 **a)** 73 km
  b) 11·6°
21 6000 m²

STAGE
10

5 Trends and time series

Exercise 5.1 (page 213)

1 a) and **c)**

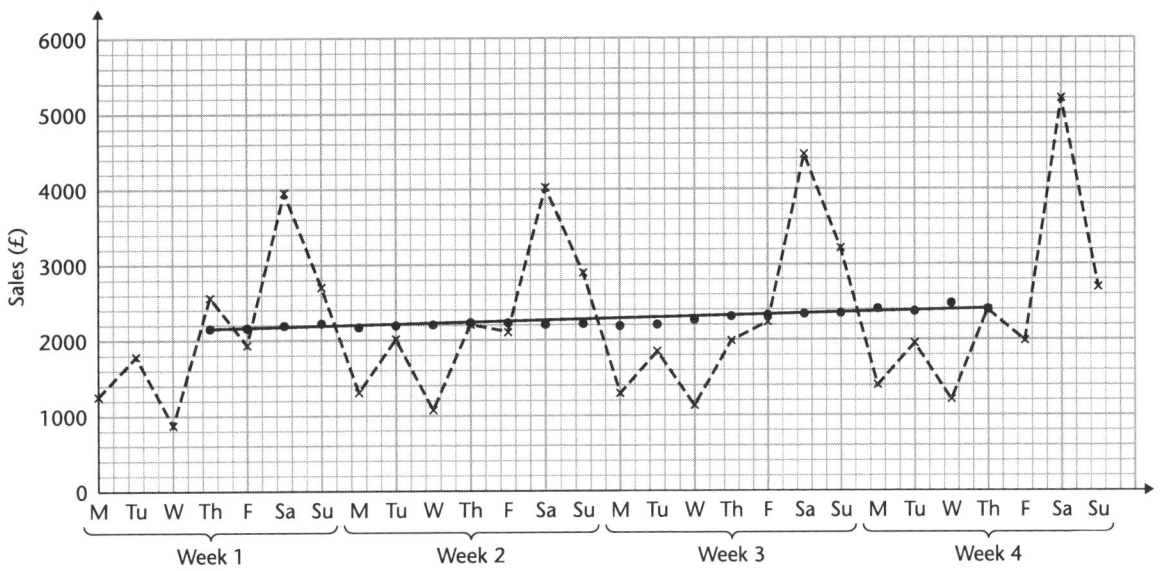

b) 2153, 2160, 2193, 2221, 2169, 2193, 2203, 2230, 2228, 2204, 2212, 2183, 2201, 2263, 2309, 2325, 2340, 2350, 2408, 2373, 2480, 2405

d) Trend is for a slight increase over the 4 weeks. Saturdays have the highest takings, Wednesdays the smallest.

2 a) Approximately £250.

b) Largest bills in 1st quarter of year. Gradual trend of decreased cost of bills.

3 a) and **c)**

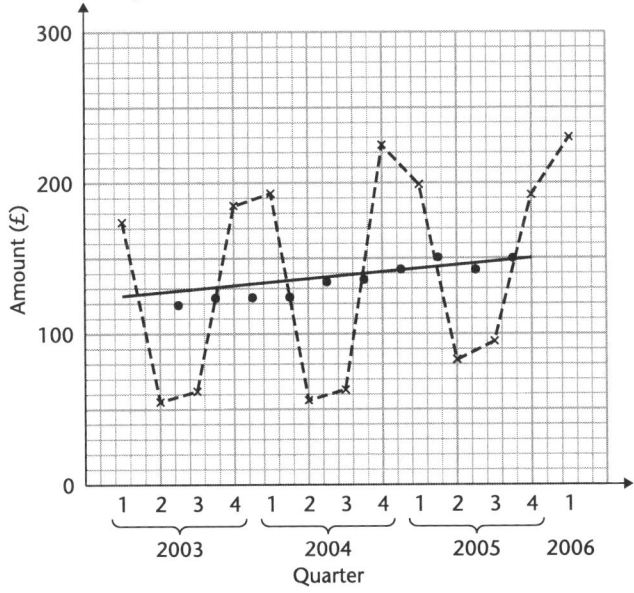

b) 119, 123·75, 124, 124·25, 134·25, 135·75, 142·5, 150·5, 142·25, 150

d) The actual values rise and fall according to the seasons; the trendline shows a steady increase.

e) The next moving average value, from the trendline, will be about £152.

$$\frac{95 + 192 + 230 + x}{4} = 152$$

So an estimate for the bill for the 2nd quarter of 2006 is £91.

4 a) and **c)**

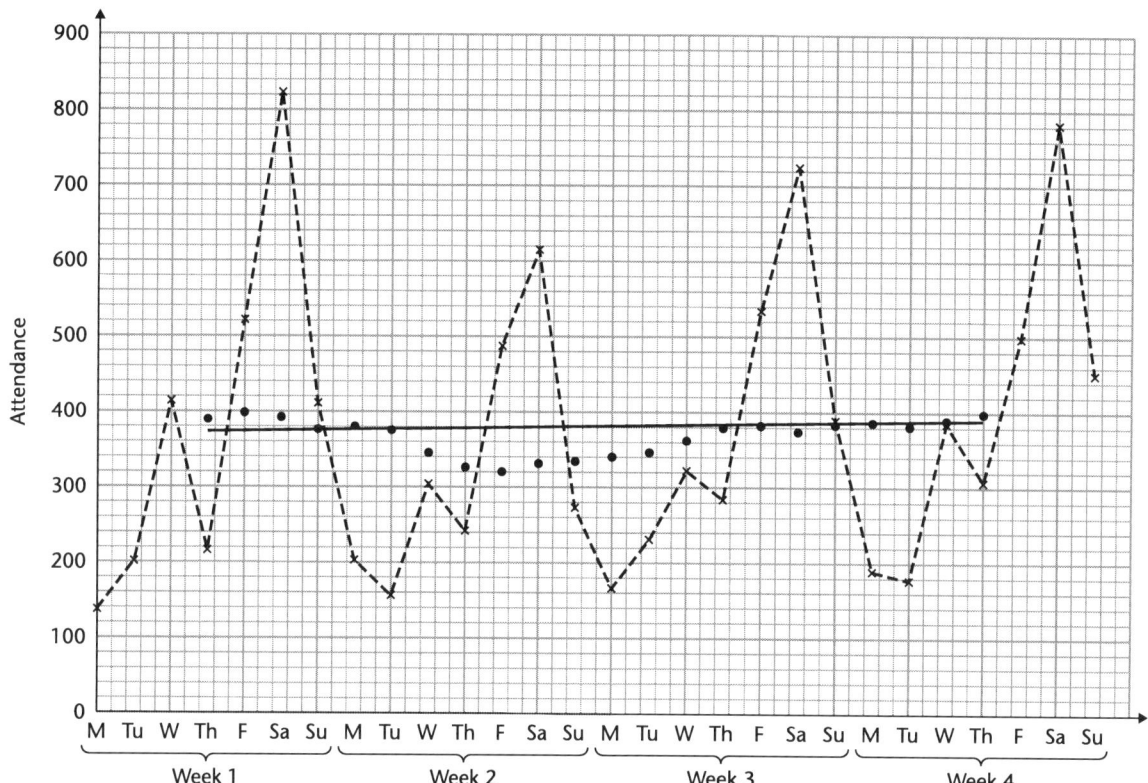

b) 389, 398, 392, 376, 380, 375, 345, 326, 320, 331, 334, 340, 346, 362, 379, 382, 374, 383, 386, 381, 389, 398

d) Trend: drop in middle of period, but recovers.
Daily variation: highest attendance on Saturdays; lower on Mondays and Tuesdays.

5 a) Overall trend downwards, but stable from 2000 to 2004.

b) 1·5–1·6%; has been going down by approximately 0·4% p.a. for last 4 years.

6

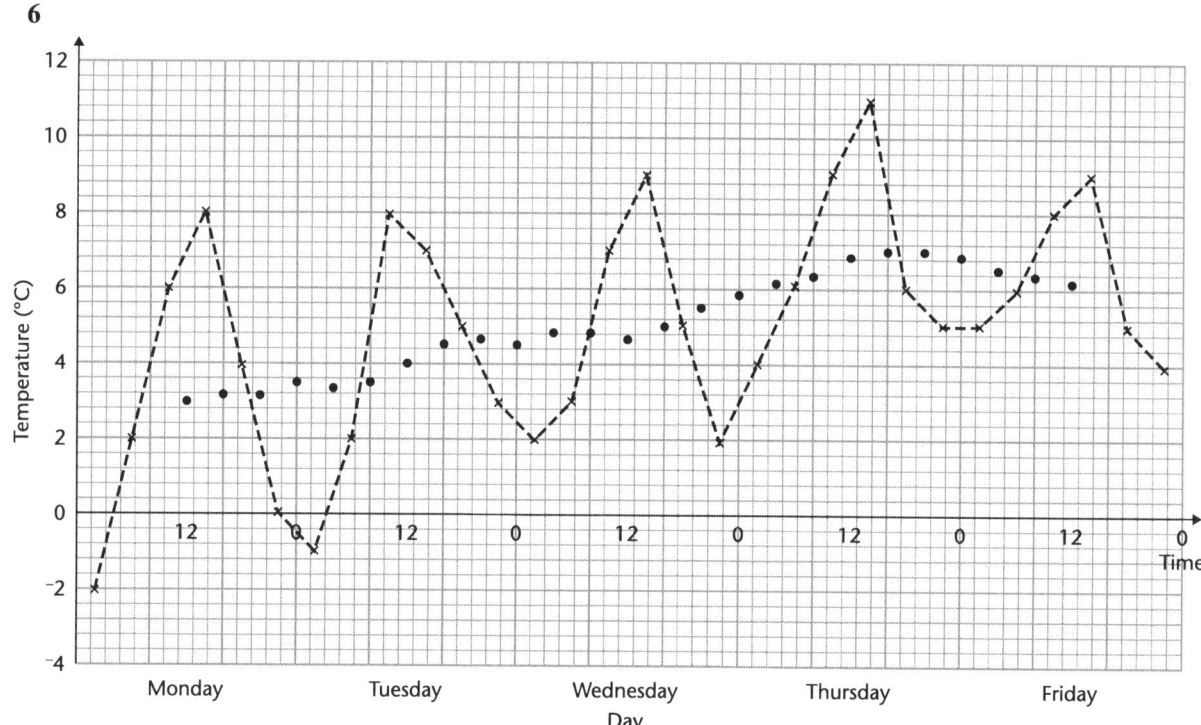

Temperature has been rising through the week, with a dip at the end.

7 a) This is the average for the last three quarters of 2004 and the first of 2005, so is $\dfrac{160 + 70 + 220 + 130}{4} = 145$. There are two moving averages which include the highest quarters (second and fourth quarters of 2004). These are for 2004 and for the last three quarters of 2004 plus the first of 2005. The latter is higher than the former since the first quarter of 2005 is higher than the first quarter of 2004.

b) The trend is for fairly constant spending on photocopying. There was a steep increase in 2004 but by the middle of 2005 spending had fallen to a level just a little above the level prior to 2004.

c) £60

d) The estimate may be poor since there is no evidence that the trend is seasonal and the trend has not been consistent over the whole time period.

8 The acceleration is positive, but is decreasing towards zero.

9 a) 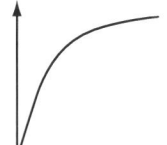   **b)** 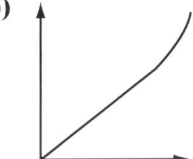   **c)** 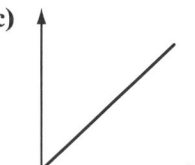   **d)**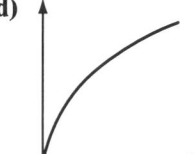

10 a) A Water going in (at steady rate).
B Tap turned to decrease water flow in.
C Tap turned to increase water flow in.
D Tap turned off.
E Plug taken out.

b) 40 litres/minute

11

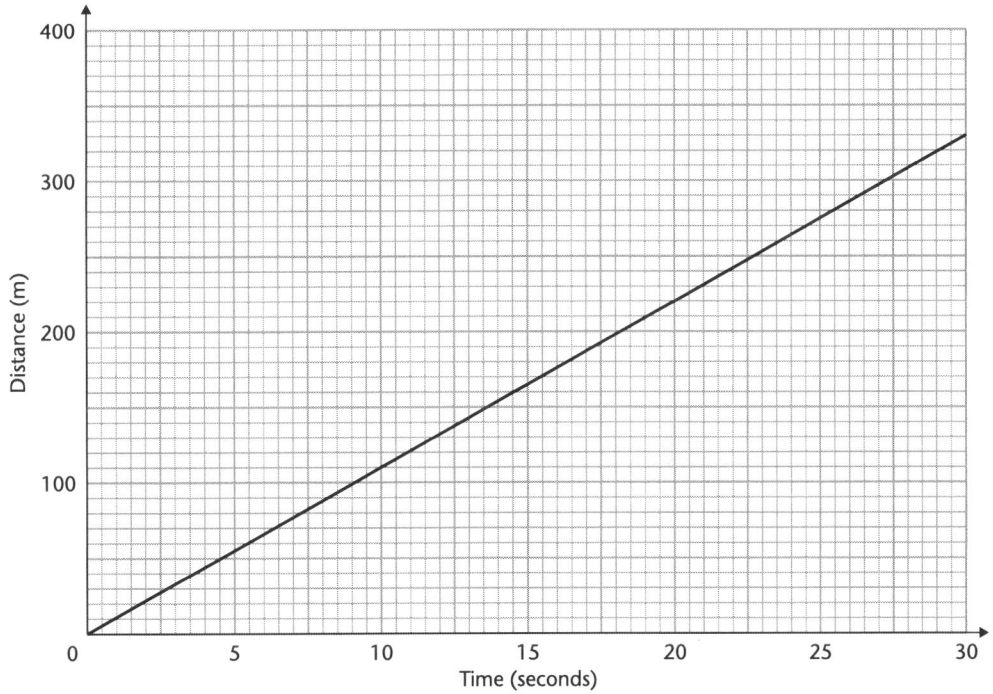

12 $6 \, \text{m/s}^2$

STAGE 10

6 Congruency – proving and using

Exercise 6.1 (page 221)

1 Triangle ABC is congruent to triangle EFD (SSS).

2 **a)** Congruent (SAS).
   b) Not congruent.
   c) Third angle = 80° therefore congruent (ASA).
   d) Congruent (RHS).
   e) Congruent (SSS).
   f) Not congruent.

3 A and (iv) (SAS)
  B and (iii) (ASA)

4 In triangles ABC and ADC:
  AB = AD (Given)
  BC = DC (Given)
  AC is common.
  So triangles ABC and ADC are congruent (SSS).
  Hence Angle BAC = Angle DAC, and so AC bisects Angle A, as required.
  Also Angle BCA = Angle DCA, and so AC bisects Angle C, as required.

5 Let D be the midpoint of BC.

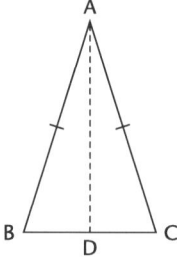

Then in triangles ABD and ACD:
AB = AC (Equal sides of isosceles triangle)
BD = CD (D is midpoint of BC)
AD is common.
So triangles ABD and ACD are congruent (SSS).
Hence Angle B = Angle C, as required.

6

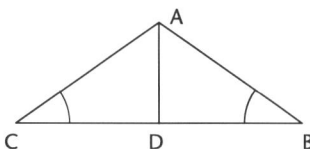

Because triangle ABC is isosceles, in triangles ACD and ABD:
AC = AB and Angle ACD = Angle ABD
CD = DB (Given)
Therefore triangle ACD is congruent to triangle ABD (SAS).

Therefore Angle CAD = Angle BAD and Angle ADC = Angle ADB.
But Angles ADC and ADB add to 180°.
Therefore they are both 90°.
Therefore AD bisects angle BAC and is perpendicular to BC.

7 ABCD is a rectangle.

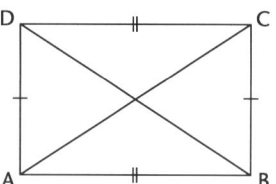

In triangles ABD, ABC:
AB is common to both, AD = BC,
Angles DAB and ABC are both 90°,
therefore triangle ABD is congruent to triangle BAC (SAS).
Therefore BD = AC.

8 ABCD is a rhombus whose diagonals cut at E.

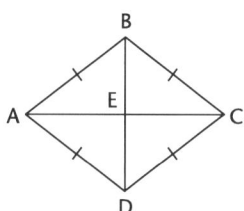

In triangles BAD and BCD:
BA = BC (ABCD is a rhombus.)
AD = CD (ABCD is a rhombus.)
BD is common.
Therefore triangles BAD and BCD are congruent (SSS).
In triangles ABE and CBE:
AB = BC (ABCD is a rhombus.)
BE is common, and because triangles BAD and BCD are congruent then Angles ABE and EBC are equal.
Therefore triangles ABE and CBE are congruent (SAS).
Therefore AE = EC and Angles AEB and BEC are equal.
Because they are on a straight line, AC, they must be 90° and this means that diagonal BD bisects AC at right angles.
Similarly you can prove the result for AC cutting BD.
Therefore the diagonals bisect each other at right angles.

9 In triangles ABD and CBD,

AB = BC ⎫ pairs of equal sides since
AD = CD ⎭ ABCD is a kite.

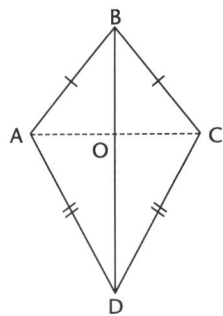

BD is common.
So triangles ABD and CBD are congruent (SSS).
This means that Angle ABD = Angle CBD and Angle ADB = Angle CDB
i.e. BD bisects each of Angles B and D.
Consider triangle ABC.
Let the diagonals intersect at O.

AB = BC ⎫
Angle ABO = Angle CBO ⎭ already shown.

BO is common.
So triangles ABO and CBO are congruent (SAS).
Hence AO = OC and Angle AOB = Angle COB = 90°
i.e. the shorter diagonal is bisected at right angles as required.

10 AX, XC, AY, YB, BZ and ZC are all equal.
The angles at A, B and C are equal.

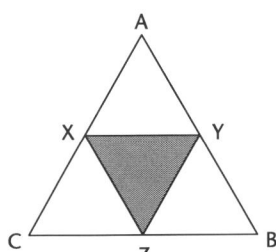

Therefore triangles CXZ, BZY and AYX are congruent (SAS).
Therefore XZ = ZY = YX.
Therefore XYZ must be an equilateral triangle.

11 The rotation about O rotates A on to A'.

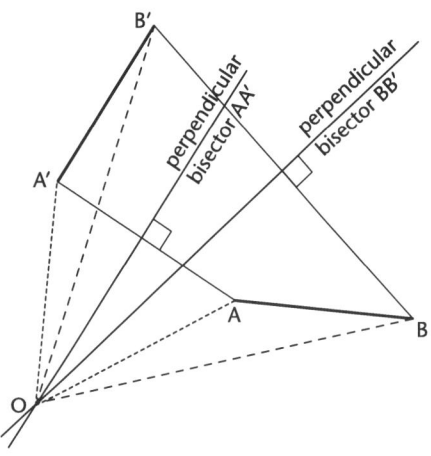

So OA = OA' and triangle OAA' is isosceles.
As shown earlier (see question **6**) the perpendicular bisector of AA' passes through O.
Similarly, the perpendicular bisector of BB' passes through O.
Hence O is the intersection of the perpendicular bisectors of AA' and BB'.
Now consider triangles OAB and OA'B'.

OA = OA' ⎫
OB = OB' ⎭ already shown.

Angle AOA' = Angle BOB' = Angle of rotation
Angle AOB = Angle A'OB' = Angle of rotation – Angle AOB'.
Hence triangles OAB and OA'B' are congruent (SAS).
Hence AB = A'B' as required.

12 Angle DAE = 40°
(Angles on a straight line add up to 180°.)
Triangle ABC is isosceles
(Since AB = AC, given.)
Angle ABC = Angle ACB = 70°
(Base angles of an isosceles triangle are equal.)
Angle ECF = 110°
(Angles on a straight line add up to 180°.)
Triangle CEF is isosceles
(Since CE = CF, given.)
Angle CEF = Angle CFE = 35°
(Base angles of an isosceles triangle are equal.)
Angle AED = 35°
(Vertically opposite angles are equal.)
Angle ADE = 105°
(Angles in a triangle add up to 180°.)

13 Angle DCB = Angle DCA
    (CD bisects Angle ACB.)
    Angle DCA = Angle CAE
    (Alternate angles.)
    Angle DCB = Angle AEC
    (Corresponding angles.)
    So Angle AEC = Angle CAE
    So triangle ACE is isosceles and AC = CE.
14 Angle ACB = Angle ABC = 70°
    (Base angles of an isosceles triangle are equal.)
    Angle ACD = 40° (Alternate angles.)
    So Angle BCD = 110°
    So Angle DBC = Angle BDC = 35°
    (Base angles of an isosceles triangle are equal.)
15 Angle BCD = 122° (Alternate angles.)
    Angle DCE = 58°
    (Angles on a straight line add up to 180°.)
    Angle BCY = 58°
    (Vertically opposite angles are equal.)
    Angle ABC = 58°
    (Angles on a straight line add up to 180°.)
    Angle BAD = 89°
    (Angle sum of triangle BAE.)
16 Angle BDC = 80°
    (Base angles of an isosceles triangle are equal.)
    So Angle DBC = 20°
    (Angles in a triangle add up to 180°.)
    So Angle ABC = 40° (BD bisects Angle ABC.)
    So Angle BAC = 60°
    (Angle sum of triangle BAC.)
17 Angle BRQ = 125°
    (Angles in a quadrilateral add up to 360°.)
    Angle BRP = 55°
    (Angles on straight line add up to 180°.)
    So a = 145°
    (Exterior angle of triangle is equal to the sum of the opposite, interior angles.)

7 Calculating the roots of equations

Exercise 7.1 (page 230)

1 x = 0·76 or 5·24
2 x = ⁻0·24 or 8·24
3 x = ⁻9·47 or ⁻0·53
4 x = ⁻3·30 or 0·30
5 x = 0·30 or 6·70
6 x = ⁻0·73 or 2·73
7 x = ⁻6·61 or 0·61
8 x = ⁻7·61 or ⁻0·39
9 x = 0·64 or 9·36
10 x = ⁻4·19 or 1·19
11 x = 0·44 or 4·56
12 x = ⁻1·18 or 0·43
13 x = 0·19 or 1·31
14 x = ⁻0·79 or 2·12

15 x = ⁻5·74 or ⁻0·26
16 x = ⁻1·22 or 0·55
17 x = ⁻0·48 or 1·68
18 x = ⁻0·35 or 2·85
19 x = ⁻0·36 or 0·56
20 x = ⁻0·65 or ⁻0·10
21 a) $(x + 6)^2 - 24$
    b) ⁻24
    c) $x = ⁻6 \pm 2\sqrt{6}$

Exercise 7.2 (page 232)

1 x = ⁻0·84 or ⁻7·16
2 x = ⁻0·82 or 1·82
3 x = ⁻1·85 or 0·18
4 x = 0·54 or 1·86
5 x = ⁻2·32 or 0·52
6 x = ⁻0·19 or 5·19
7 x = ⁻2·26 or ⁻0·74
8 x = ⁻0·63 or ⁻6·37
9 x = ⁻0·85 or 2·35
10 x = ⁻1·22 or 0·55
11 x = 0·76 or 1·84
12 x = ⁻0·44 or ⁻1·36
13 x = ⁻0·16 or 0·88
14 x = ⁻1·90 or 1·23
15 x = ⁻5·74 or ⁻0·26
16 x = ⁻1·22 or 0·55
17 x = ⁻0·48 or 1·68
18 x = ⁻0·35 or 2·85
19 x = ⁻0·36 or 0·56
20 x = ⁻0·65 or ⁻0·10
21 Width 2·40 m, length 10·40 m
22 15 and 45 years
23 a) $20x - 2x^2$
    b) 7·24 m by 5·53 m or 2·76 m by 14·47 m
    c) 50 m²
24 0·90 m

8 Surface areas and complex shapes

Exercise 8.1 (page 237)

1 a) 242 cm²   b) 18·8 m²
   c) 101 cm²   d) 57·7 cm²
   e) 19·1 m²   f) 710 cm²
2 a) 92·4 cm²   b) 204 cm²
   c) 58·0 cm²   d) 121 cm²
   e) 427 cm²   f) 135 cm²
3 a) 314 cm²   b) 483 cm²
   c) 50·3 mm²   d) 113 cm²
   e) 278 cm²   f) 765 mm²
4 61·3 cm²
5 1·95 cm
6 173 cm²
7 375 cm²

8 5420 cm²

9 255 cm²

10 7·6 cm

11 47·7 cm

12 153 cm²

13 170 cm²

14 3·54 cm

15 4·3 cm

16 75·2 cm²

17 130 cm²

Exercise 8.2 (page 241)

1 63·3 cm²

2 a) 5·74 cm   **b)** 11·8 cm

3 a) 3 cm   **b)** 1225 cm³

4 Check students' proof.

5 a) 3·54 cm   **b)** 29·5 cm³

6 a) Check students' proof.

   b) 6·13 cm

7 a) 3·34 cm   **b)** 72·0 cm²

8 484 cm²

9 38·4 cm²

10 169 cm³

11 24·0 cm²

12 29·6 cm

13 8·36 cm

14 a) Check students' proof.

   b) 218 cm³

15 204 cm²

16 2·84 cm

17 72·3 cm²

18 124 cm³

19 a) Check students' proof.

   b) 1619 cm²

20 81·4 cm³

21 a) 409 cm³   **b)** 10·7 cm, 382 cm²

22 a) Check students' proof.

   b) 2·56 litres

23 1740 cm³

24 4·8 cm, 174 cm³

25 784 cm³

Revision exercise B1 (page 246)

1 a) and **b)**

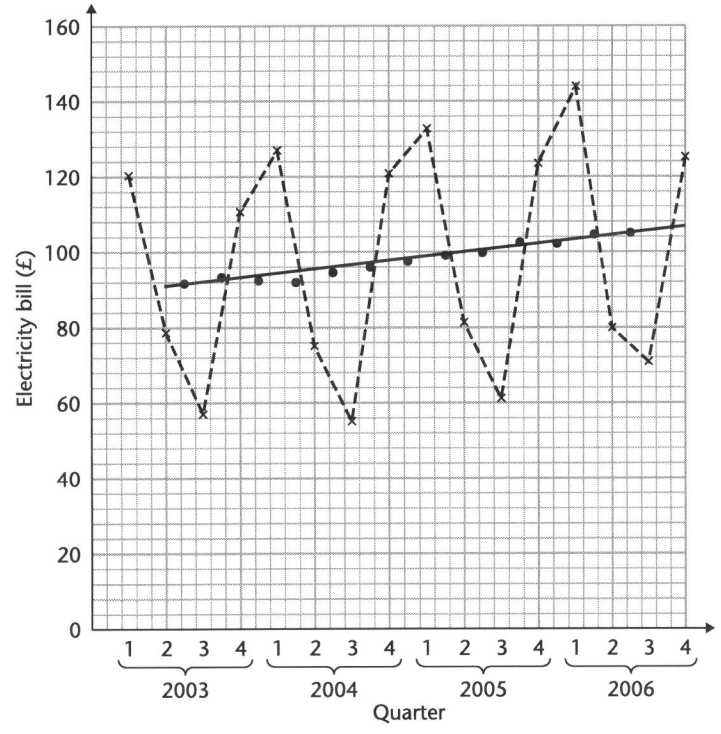

c) A slight upward trend.

d) Possible estimates are £145 – £150, £80 – £83, £65 – £75, £125 – £130.

STAGE
10

2 a) Rainy: Nov – May

Dry: June – Oct

b)

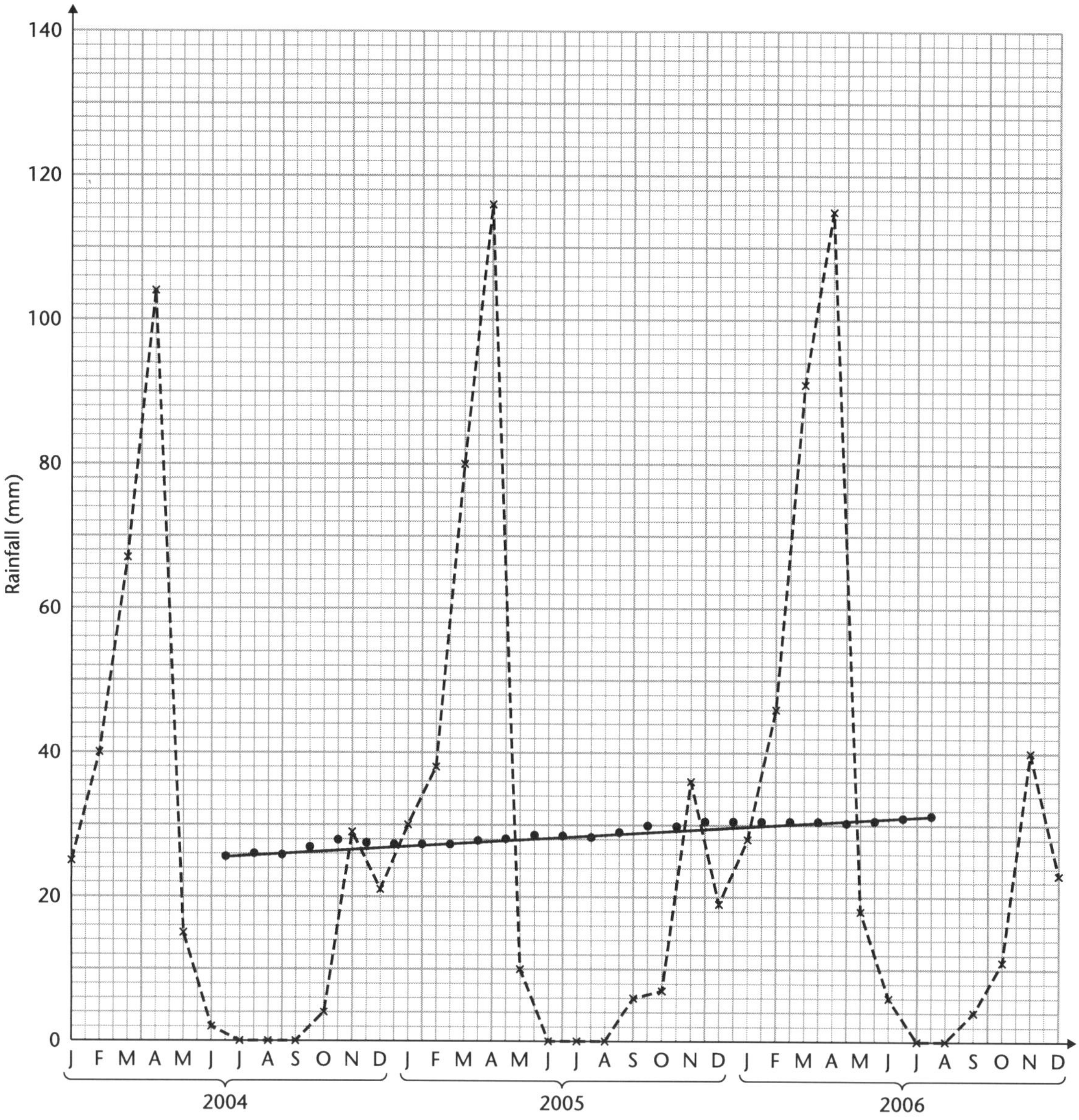

The graph shows a slight increase in rainfall over the 3-year period.

3 a) A The oven is warming up at a constant rate.

B The door is opened as the oven is loaded and the temperature starts to fall.

C The door is shut again.

D The door is opened at the end of baking (the oven is turned off).

b) Approximately 12°C per minute.

4 Because joining the midpoints as shown creates isosceles triangles, all the marked angles are 45°. Therefore the angles of the quadrilateral are 90°. The triangles ZAW, WBX, XCY and YDZ are congruent (SAS), therefore the sides ZW, WX, XY and YZ are equal. Therefore ZWXY is a square.

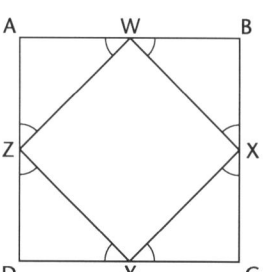

5 OA = OB (Radii)
   ∠OAP = ∠OAB = 90° (Angle between tangent and radius.)
   OP is common to both triangles.
   So triangles OAP and OBP are congruent (RHS).
6 PA = PB (Tangents from a point to a circle are equal.)
   ∠APD = ∠DPB (Triangles OAP and OBP are congruent.)
   DP is common.
   So triangles PAD and PBD are congruent (SAS).
7 a) $x = 0.35$ or 5.65
   b) $x = {}^-0.44$ or 1.69
   c) $x = {}^-1.72$ or 0.39
   d) $x = 1.82$ or 14.82
   e) $x = 1.47$ or $^-6.47$
8 $y = (x - 2.5)^2 - 2.25$; $(2.5, {}^-2.25)$
9 a) Area of lawn = area of path
       $3x^2 = (x + 1)(3x + 2) - 3x^2$
       $3x^2 = 3x^2 + 5x + 2 - 3x^2$
       $3x^2 = 5x + 2$
   b) Length = 2 m, width = 6 m
10 a) $x = 0.137$ or 1.463
   b) $x = {}^-0.275$ or 7.275
   c) $x = {}^-1.260$ or 0.926
   d) $x = 1.174$ or $^-2.840$
   e) $x = {}^-1.243$ or 0.643
11 $18.7\,\text{cm}^2$
12 a) 12 cm   b) $151\,\text{cm}^2$
13 $65.6\,\text{cm}^2$
14 44

9 Working with algebraic fractions

Exercise 9.1 (page 251)

1 $\dfrac{11x}{10}$

2 $\dfrac{x}{15}$

3 $\dfrac{^-4x + 5}{6}$

4 $\dfrac{3x + 1}{10}$

5 $\dfrac{17x - 12}{30}$

6 x

7 $\dfrac{3x - 1}{x(x - 1)}$

8 $\dfrac{5x + 3}{x(x + 1)}$

9 $\dfrac{x + 6}{2x(x + 2)}$

10 $\dfrac{4x + 5}{6x(2x + 1)}$

11 $\dfrac{5x + 1}{(x + 1)(x - 1)}$

12 $\dfrac{8x + 7}{(x + 2)(x - 1)}$

13 $\dfrac{2x^2 - 9x - 5}{(3x + 1)(x + 3)}$ or $\dfrac{(2x + 1)(x - 5)}{(3x + 1)(x + 3)}$

14 $\dfrac{x^2 + 6x + 1}{(x + 1)(x + 3)}$

15 $\dfrac{4x^2 - x + 3}{(x - 1)(x + 2)}$

16 $\dfrac{^-x^2 + 5x + 2}{(x - 1)(x + 2)}$

17 $\dfrac{7x^2 - 8x - 10}{5x(x + 1)}$

18 $\dfrac{3x^2 - 17x - 15}{5x(x + 1)}$

19 $\dfrac{23x^2 + 4x + 51}{9(x - 3)(x + 2)}$

20 $\dfrac{^-2x^2 + 3x + 23}{(x - 1)(x + 2)(x + 3)}$

21 $\dfrac{3(x + 1)(2x + 3)}{(2x + 1)(x + 2)}$

22 $\dfrac{2(x^2 + 7x - 3)}{(x + 3)(x - 3)}$

Exercise 9.2 (page 254)

1 $x = 5$
2 $x = 2$
3 $x = 3$
4 $x = 3$
5 $x = 0$
6 $x = 4$
7 $x = {}^-3$
8 $x = 5$
9 $x = 1$
10 $x = \frac{^-2}{3}$
11 $x = {}^-2$ or 2
12 $x = {}^-3$ or $^-4$
13 $x = {}^-1$ or 5
14 $x = {}^-2$ or 5
15 $x = {}^-1$ or $\frac{3}{4}$
16 $x = 1$ or $\frac{3}{4}$
17 $x = \frac{^-1}{2}$ or 4
18 $x = \frac{3}{2}$ or $\frac{^-5}{3}$
19 $x = {}^-11$ or 2
20 $x = \frac{^-1}{2}$ or 5
21 $x = 5.303$ or 1.697
22 $x = 0.463$ or $^-0.863$
23 $x = 2.303$ or $^-1.303$
24 $x = {}^-0.21$ or 5.96

10 Vectors

Exercise 10.1 (page 259)

1 $\overrightarrow{AB} = \binom{4}{1}$, $\overrightarrow{CD} = \binom{2}{0}$, $\overrightarrow{CB} = \binom{1}{4}$, $\overrightarrow{AD} = \binom{5}{-3}$,

$\overrightarrow{CA} = \binom{-3}{3}$

2 $\overrightarrow{EF} = \binom{4}{3}$, $\overrightarrow{GH} = \binom{-2}{-2}$, $\overrightarrow{EH} = \binom{6}{1}$, $\overrightarrow{GF} = \binom{-4}{0}$,

$\overrightarrow{FH} = \binom{2}{-2}$

3 $\overrightarrow{AB} = \binom{1}{3}$, $\overrightarrow{CD} = \binom{0}{-2}$, $\overrightarrow{CB} = \binom{-4}{2}$, $\overrightarrow{AD} = \binom{5}{-1}$,

$\overrightarrow{CA} = \binom{-5}{-1}$

4 $\overrightarrow{AB} = \binom{6}{1}$, $\overrightarrow{CD} = \binom{-6}{-2}$, $\overrightarrow{CB} = \binom{-2}{-5}$, $\overrightarrow{AD} = \binom{2}{4}$,

$\overrightarrow{CA} = \binom{-8}{-6}$

5 a) $\binom{0}{2}$ b) $\binom{-4}{0}$

   c) $\binom{-2}{3}$ d) $\binom{1}{7}$

   e) $\binom{8}{-6}$ f) $\binom{-6}{4}$

6 $\overrightarrow{AB} = 2\mathbf{a}$, $\overrightarrow{CD} = {}^{-}4\mathbf{a}$, $\overrightarrow{EF} = 3\mathbf{a}$, $\overrightarrow{GH} = {}^{-}2\mathbf{a}$,

$\overrightarrow{PQ} = {}^{-}\tfrac{1}{2}\mathbf{a}$, $\overrightarrow{RS} = \tfrac{7}{2}\mathbf{a}$

7 a) (4, 4)   b) (6, 4)
   c) (−2, 2)   d) (4, −1)
   e) (−8, 0)   f) (0, 0)
   g) (5, 6)   h) (8, 1)
   i) (2, 1)   j) (2, 1)
   k) (1, −5)   l) (8, −6)

8 $\overrightarrow{AB} = 2\mathbf{a}$, $\overrightarrow{CD} = {}^{-}\mathbf{a}$, $\overrightarrow{EF} = \tfrac{1}{2}\mathbf{a}$, $\overrightarrow{GH} = \tfrac{3}{2}\mathbf{a}$,

$\overrightarrow{PQ} = {}^{-}\tfrac{1}{2}\mathbf{a}$, $\overrightarrow{RS} = \tfrac{9}{4}\mathbf{a}$

9 $\overrightarrow{AB} = \mathbf{a}$, $\overrightarrow{CD} = {}^{-}\mathbf{b}$, $\overrightarrow{EF} = 2\mathbf{b}$, $\overrightarrow{GH} = {}^{-}\tfrac{1}{2}\mathbf{a}$,

$\overrightarrow{PQ} = {}^{-}\tfrac{1}{2}\mathbf{b}$, $\overrightarrow{RS} = 3\mathbf{a}$

10 $\overrightarrow{AB} = \mathbf{b}$, $\overrightarrow{CD} = {}^{-}\mathbf{a}$, $\overrightarrow{EF} = 2\mathbf{a}$, $\overrightarrow{GH} = 2\mathbf{b}$,

$\overrightarrow{PQ} = \tfrac{1}{2}\mathbf{a}$, $\overrightarrow{RS} = \tfrac{1}{2}\mathbf{b}$

11 $\overrightarrow{AB} = 2\mathbf{a}$, $\overrightarrow{CD} = {}^{-}2\mathbf{a}$, $\overrightarrow{EB} = \mathbf{a}$,

$\overrightarrow{GD} = {}^{-}\mathbf{a}$, $\overrightarrow{HF} = 2\mathbf{a}$, $\overrightarrow{FC} = \mathbf{b}$

12 $\overrightarrow{BC} = \mathbf{b}$, $\overrightarrow{CD} = {}^{-}\mathbf{a}$, $\overrightarrow{EB} = \tfrac{1}{2}\mathbf{a}$, $\overrightarrow{HD} = \tfrac{1}{2}\mathbf{b}$,

$\overrightarrow{HF} = \mathbf{a}$, $\overrightarrow{FB} = {}^{-}\tfrac{1}{2}\mathbf{b}$

Exercise 10.2 (page 266)

1 a) $\binom{4}{6}$ b) $\binom{9}{3}$ c) $\binom{2}{3}$

  d) $\binom{1}{0}$ e) $\binom{5}{12}$

2 a) $\binom{-6}{0}$ b) $\binom{-1}{-2}$ c) $\binom{0.5}{-1.5}$

  d) $\binom{6}{1}$ e) $\binom{0}{1}$

3 a) $\binom{3}{12}$ b) $\binom{8}{12}$ c) $\binom{4}{5}$

  d) $\binom{7}{4}$ e) $\binom{7}{18}$

4 a) $\binom{-2}{0}$ b) $\binom{-6}{3}$ c) $\binom{-1}{2}$

  d) $\binom{-3}{-10}$ e) $\binom{-0.5}{5.5}$

5 a) $\binom{12}{6}$ b) $\binom{-6}{-3}$ c) $\binom{24}{12}$

  d) $\binom{3}{1.5}$ e) $\binom{-2}{-1}$

6 a) $\binom{3}{9}$ b) $\binom{4}{7}$ c) $\binom{2}{1}$

  d) $\binom{5}{10}$ e) $\binom{-3}{1}$

7 a) $\binom{-3}{-9}$ b) $\binom{2}{-20}$ c) $\binom{4}{-4}$

  d) $\binom{-7}{12}$ e) $\binom{4.5}{-1}$

8 a) $\binom{20}{32}$ b) $\binom{-10}{-16}$ c) $\binom{2.5}{4}$

  d) $\binom{45}{72}$ e) $\binom{2}{3.2}$

9 a) $\binom{8}{2}$ b) $\binom{9}{4}$ c) $\binom{1}{2}$

  d) $\binom{13}{5}$ e) $\binom{7}{7}$

10 a) $\binom{-6}{-9}$ b) $\binom{0}{1}$ c) $\binom{-7}{-4}$

  d) $\binom{14}{30}$ e) $\binom{3}{8.5}$

Exercise 10.3 (page 269)

1

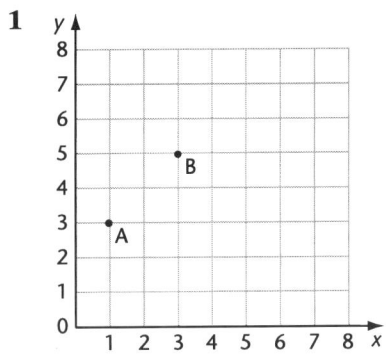

a) $\begin{pmatrix} 2 \\ 2 \end{pmatrix}$

b) **(i)** $\begin{pmatrix} 2 \\ 2 \end{pmatrix}$ **(ii)** $\begin{pmatrix} 2 \\ 2 \end{pmatrix}$ **(iii)** $\begin{pmatrix} 2 \\ 2 \end{pmatrix}$

c) They are all $\begin{pmatrix} 2 \\ 2 \end{pmatrix}$.

2

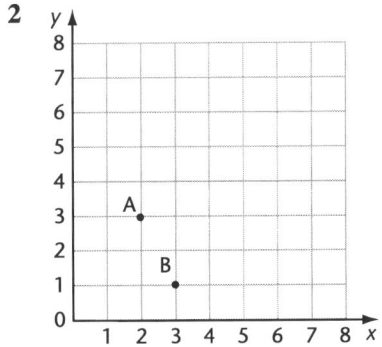

a) $\overrightarrow{AB} = \begin{pmatrix} 1 \\ -2 \end{pmatrix}$

b) **(i)** $\begin{pmatrix} 1 \\ -2 \end{pmatrix}$ **(ii)** $\begin{pmatrix} 1 \\ -2 \end{pmatrix}$ **(iii)** $\begin{pmatrix} 1 \\ -2 \end{pmatrix}$

c) They are all the same vector.

3

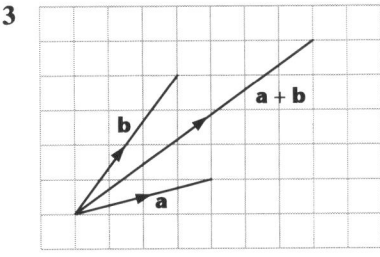

4

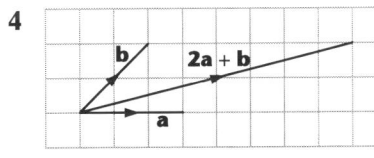

5 a) **(i)** $\overrightarrow{AB} = \begin{pmatrix} 6 \\ 2 \end{pmatrix}$ **(ii)** $\overrightarrow{BC} = \begin{pmatrix} 3 \\ 1 \end{pmatrix}$

    b) $\overrightarrow{AB} = 2 \times \overrightarrow{BC}$. So ABC is a straight line and AB = 2 × BC in length.

6 a) **(i)** $\overrightarrow{AB} = \begin{pmatrix} 2 \\ 3 \end{pmatrix}$ **(ii)** $\overrightarrow{CD} = \begin{pmatrix} -4 \\ -6 \end{pmatrix}$

    b) The line AB is parallel to DC and CD = 2 × AB in length.

7 $\overrightarrow{BC} = \overrightarrow{BA} + \overrightarrow{AC} = {}^-\mathbf{a} + 2\mathbf{b} = 2\mathbf{b} - \mathbf{a}$

8 $\overrightarrow{AC} = 2\mathbf{a} - 3\mathbf{b}$

9 a) $\overrightarrow{AB} = {}^-\mathbf{a} - \mathbf{b}$ **b)** $\overrightarrow{BC} = 3\mathbf{b} - 4\mathbf{a}$

    c) $\overrightarrow{AC} = 2\mathbf{b} - 5\mathbf{a}$

10 $\overrightarrow{BC} = \mathbf{b}, \overrightarrow{CD} = {}^-\mathbf{a}, \overrightarrow{BD} = \mathbf{b} - \mathbf{a}$, and $\overrightarrow{AC} = \mathbf{a} + \mathbf{b}$

11 $\overrightarrow{AB} = \mathbf{b} - \mathbf{a}, \overrightarrow{CB} = \frac{1}{3}(\mathbf{b} - \mathbf{a}), \overrightarrow{OC} = \frac{1}{3}\mathbf{a} + \frac{2}{3}\mathbf{b}$

12 $\overrightarrow{EB} = \frac{1}{2}\mathbf{a} - \mathbf{b}$

13 $\overrightarrow{CD} = \frac{1}{2}\mathbf{b} - \mathbf{a}$

14 a) $\overrightarrow{AE} = 3\mathbf{a}, \overrightarrow{AF} = 3\mathbf{b}, \overrightarrow{BC} = \mathbf{b} - \mathbf{a}$,
      $\overrightarrow{EF} = 3\mathbf{b} - 3\mathbf{a}$

    b) $\overrightarrow{EF} = 3 \times \overrightarrow{BC}$ so EF and BC are parallel and EF = 3 × BC in length.

15 a) $\overrightarrow{AB} = 3\mathbf{b}$ and $\overrightarrow{BC} = {}^-2\mathbf{a}$

    b) OABC is a parallelogram (opposite sides equal and parallel).

16 $\overrightarrow{AB} = \mathbf{a} + \mathbf{b}$

17 $\overrightarrow{EB} = \mathbf{b} - \frac{2}{3}\mathbf{a}$

18 a) $\overrightarrow{BD} = \mathbf{b} - \mathbf{a}, \overrightarrow{BE} = \frac{1}{2}(\mathbf{b} - \mathbf{a}), \overrightarrow{AE} = \frac{1}{2}(\mathbf{a} + \mathbf{b})$,
      $\overrightarrow{EC} = \frac{3}{2}(\mathbf{a} + \mathbf{b}), \overrightarrow{BC} = \mathbf{a} + 2\mathbf{b}$

    b) AD and BC are not parallel because one vector is not a multiple of the other.

19 a) $\overrightarrow{AB} = \mathbf{b} - \mathbf{a}, \overrightarrow{OC} = 4\mathbf{a}, \overrightarrow{OD} = 4\mathbf{b}$,
      $\overrightarrow{CD} = 4(\mathbf{b} - \mathbf{a})$
      As $\overrightarrow{CD}$ is a multiple of $\overrightarrow{AB}$, AB and CD are parallel.

    b) The ratio of the lengths of AB and CD is $1:4$.

20 a) $\mathbf{b}$ **b)** ${}^-\mathbf{b} - \mathbf{a}$

    c) ${}^-\mathbf{a} + \mathbf{b}$ **d)** ${}^-2\mathbf{a} + \mathbf{b}$

21 a) ${}^-\mathbf{a} + \mathbf{b}$ or $\mathbf{b} - \mathbf{a}$

    b) $\frac{1}{3}({}^-\mathbf{a} + \mathbf{b})$ or $\frac{-1}{3}(\mathbf{b} - \mathbf{a})$

    c) $\mathbf{a} + \frac{1}{3}(\mathbf{b} - \mathbf{a})$ or $\frac{2}{3}\mathbf{a} + \frac{1}{3}\mathbf{b}$

22 a) **(i)** ${}^-\mathbf{a} + \mathbf{b}$ **(ii)** ${}^-2\mathbf{a} + 2\mathbf{b}$

    b) AB is parallel to CD and half the length.

23 a) **(i)** $2\mathbf{a} + \mathbf{c}$ **(ii)** ${}^-\mathbf{a} + \mathbf{c}$
      (iii) $\frac{2}{3}\mathbf{a} + \frac{1}{3}\mathbf{c}$

    b) O, F and E are in a straight line. OE = 3 × OF.

STAGE
10

11 Comparing sets of data

Exercise 11.1 (page 278)

1

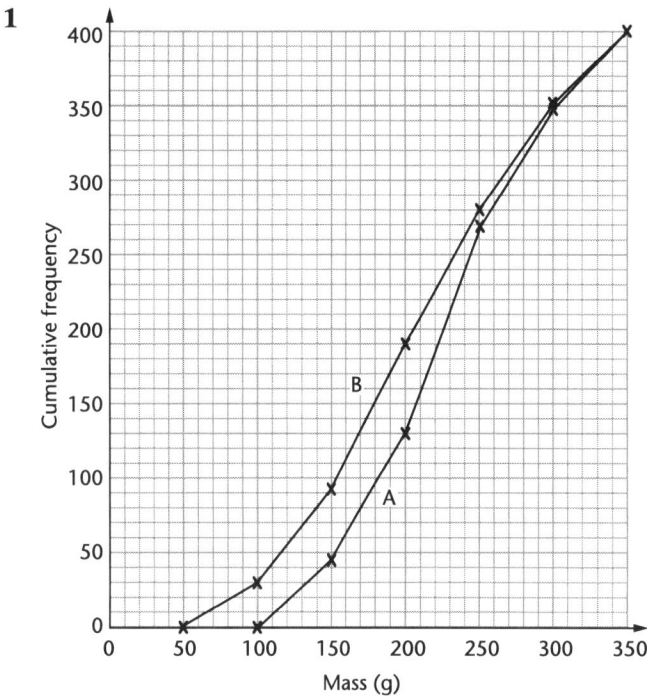

Variety B is more varied in size (spread is greater).

2 a)

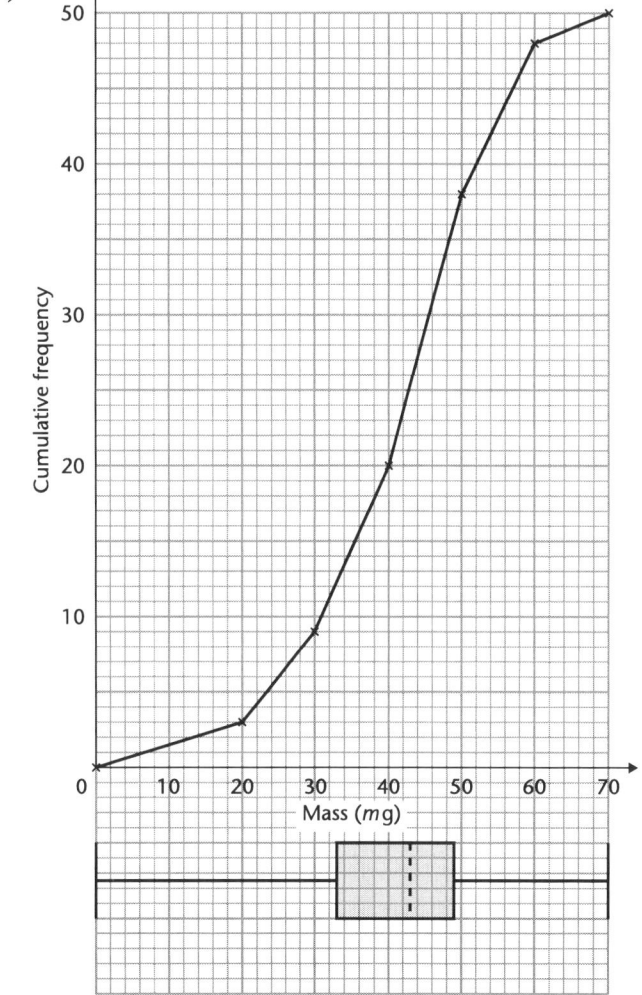

b) Sample B is smaller and so the statistics for this variety are less reliable.
Variety B is heavier on average (median for A = 43·1 g).
The masses for variety A are less spread out than those for B (IQR for A = 17·8 g, IQR for B = 25 g).

Graduated Assessment for OCR GCSE Mathematics © Hodder Murray 2007

3 a) Guildford had the highest recorded temperature, but Torquay was hotter on average (greater median).

  b) Guildford had the greater variability – both the range and interquartile range were greater.

4 a) 2160

  b)

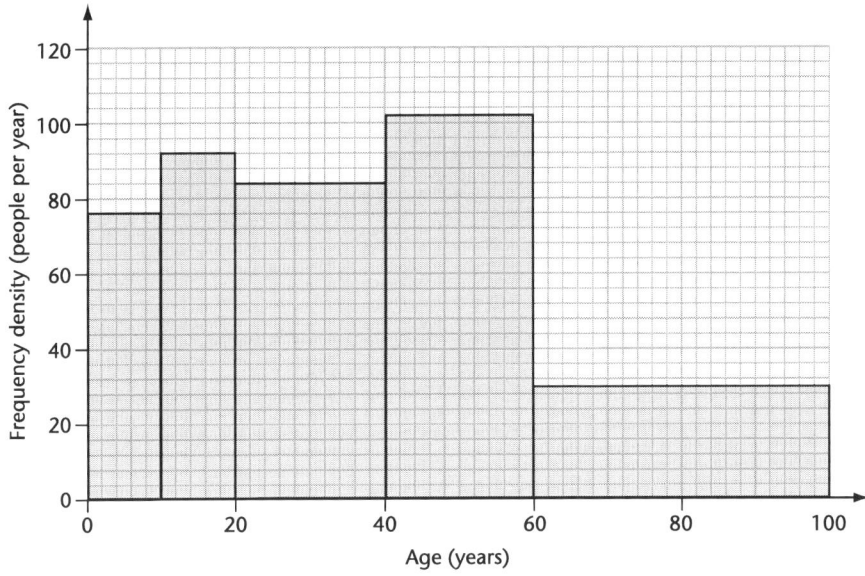

  c) Fewer people live in Banton (6600) than Aveford (7520); the same number of over 60s live in each place; a larger percentage of the population is under 10 in Aveford (16·0%) than Banton (11·5%).

5

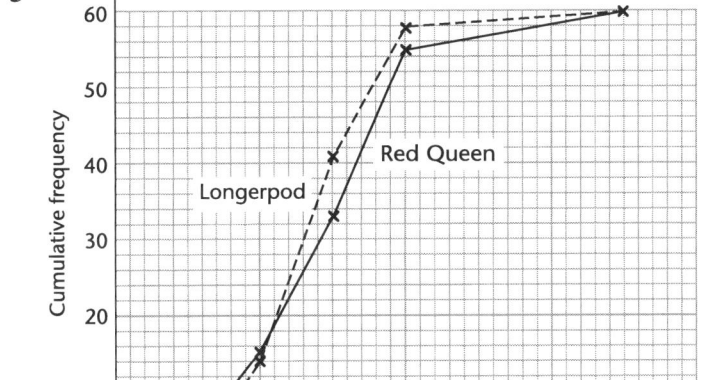

Longerpod beans are smaller on average and are more consistent in size than Red Queen.

6 Joanne: Median 12·5 minutes, LQ 5·7 minutes, UQ 17·6 minutes, IQR 11·9 minutes
  Simon: Median 7·5 minutes, LQ 3·2 minutes, UQ 12·5 minutes, IQR 9·3 minutes
  Joanne's calls are longer on average; Joanne's calls vary more in length.

7 a) That 15 children in each class have £4 or less to spend.

  b) On average, those in class B have less to spend; there are more children in A than in B.

8 a) The Bahamas plane has a greater proportion of older (over 50s) people; the Majorca plane has proportionately lots more 20–30s.

  b)

Age (a years)	Frequency Bahamas	Frequency Majorca
$0 \leqslant a < 10$	3	12
$10 \leqslant a < 20$	10	16
$20 \leqslant a < 30$	21	47
$30 \leqslant a < 50$	66	80
$50 \leqslant a < 70$	44	52
$70 \leqslant a < 80$	12	2

Bahamas 41 years, Majorca 38 years

STAGE
10

9 Mean mark is lower in English; less spread of marks in English.

10 Means: lilac 84·8 mm
         lime 81·4 mm
Lime leaves have a smaller median and range.

11 The first year group had a mean height of 167·0 cm, the second a mean of 159·3 cm, so are 7·7 cm shorter, on average.

12 a) Wellfit 233, Superhealth 222.
  b) Wellfit has a greater proportion of under 18s, Superhealth has a greater proportion of over 60s; the mean age of Wellfit's members is lower.

13 a)

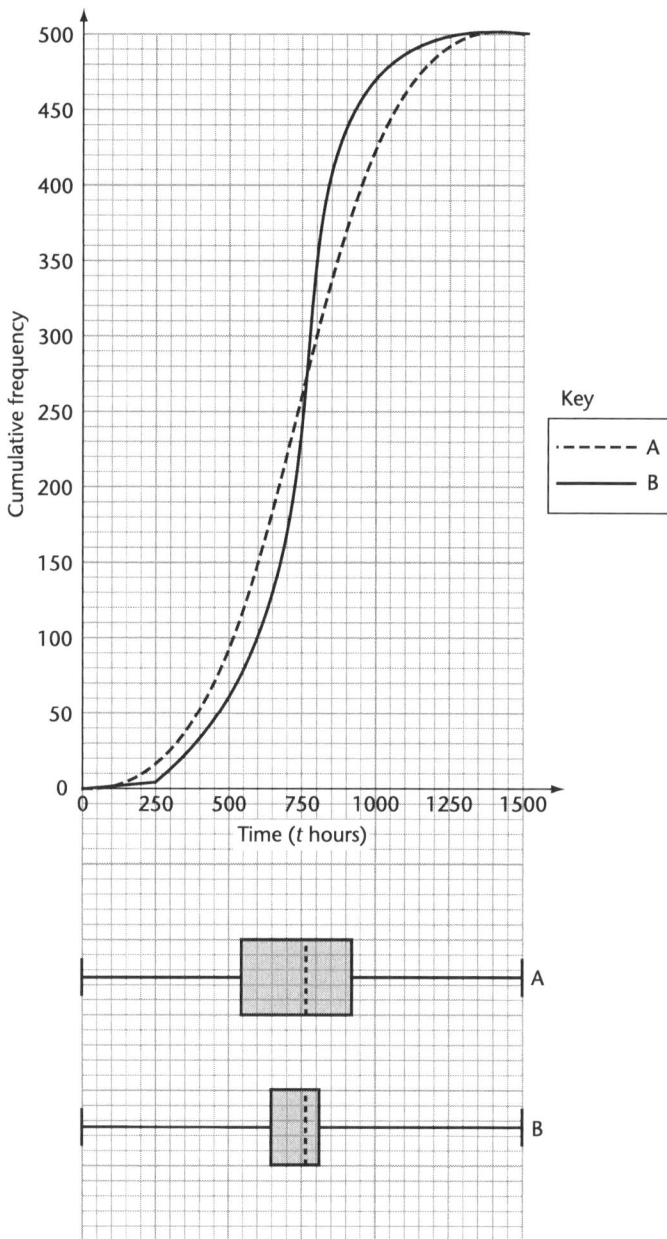

Key
- - - - - A
———— B

  b) Both types have about the same average lifetime but type B is more reliable since its interquartile range is smaller.

14 a) Translated 4 cm to right.

b)

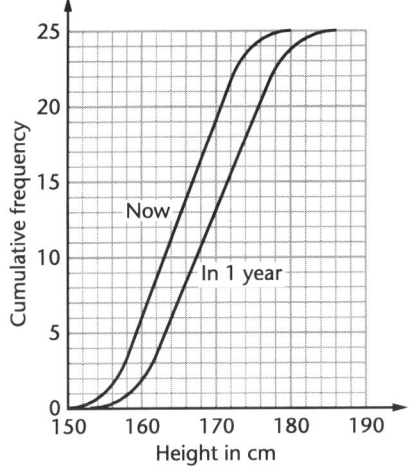

15 a) 5 girls and 18 boys
b) In general, boys spent less time on the phone; the times for the girls were more evenly spread than those for the boys; there were 50 girls and 50 boys represented.

16 a) The 1967 outbreak had nearly 2500 cases altogether and lasted 16 weeks. The number of new cases, 81 one day, were at their greatest only 5 weeks after the first case of the disease. During the last five weeks of the outbreak, there were less than 10 new cases each day.
b) Since 25 February the 2001 outbreak had fewer new cases each day compared with 1967, until 31 March when it peaked at 59 cases. To date, there are less than half the total cases there were in 1967.

12 Simultaneous equations

Exercise 12.1 (page 292)

1 $x = 2, y = 3$
2 $x = 2, y = 3$
3 $x = 1, y = 2$
4 $x = 5, y = 6$
5 $x = 2, y = 1$
6 $x = 1\frac{1}{2}, y = 1\frac{1}{2}$
7 $x = 2, y = -1$
8 $x = 2, y = 1$

Exercise 12.2 (page 294)

1 $x = 4, y = 2$ or $x = -1, y = 12$
2 $x = -1, y = 3$ or $x = 8, y = 39$
3 $x = -1, y = 5$ or $x = \frac{1}{2}, y = 2\frac{3}{4}$
4 $x = -2, y = 5$ or $x = \frac{1}{2}, y = 3\frac{3}{4}$
5 $x = 5, y = -11$ or $x = -2, y = 10$
6 $x = 2, y = 1$ or $x = 4, y = 5$
7 $x = -2, y = 5$ or $x = -1, y = 3$

8 $x = \frac{4}{5}, y = -\frac{14}{25}$ or $x = -2, y = 0$
9 $x = -1, y = 11$ or $x = 4, y = 1$
10 $x = -3, y = 17$ or $x = 2, y = -3$
11 $x = -1, y = 4$ or $x = 4, y = 19$
12 $x = -1, y = 9$ or $x = 2\frac{1}{2}, y = -3\frac{1}{4}$
13 $x = -3, y = 6$ or $x = 1, y = 2$
14 $x = 0.839, y = 3.516$ or $x = -0.239, y = 0.284$
15 $x = 0, y = 0$ or $x = 1, y = 1$

Exercise 12.3 (page 296)

1 $x = 0, y = 7$ or $x = 7, y = 0$
2 $x = 5, y = 12$ or $x = -12, y = -5$
3 $x = 0, y = 5$ or $x = 5, y = 0$
4 $x = 6, y = 8$ or $x = -8, y = -6$
5 $x = 0, y = 8$ or $x = -6.4, y = -4.8$
6 $x = 2, y = 0$ or $x = -1, y = 3$
7 $x = 9, y = 12$ or $x = -12, y = -9$
8 $x = 0, y = 3$ or $x = 3, y = 0$
9 $x = 6, y = 8$ or $x = 8, y = 6$
10 $x = -3, y = -5$ or $x = 5, y = 3$
11 $x = -3, y = 4$ or $x = -4, y = 3$
12 $x = 8, y = 6$ or $x = -6, y = -8$
13 $x = 1.41, y = 1.41$ or $x = -1.41, y = -1.41$
14 $x = 0.22, y = 2.22$ or $x = 2.22, y = -0.22$
15 $x = 3.08, y = 5.15$ or $x = -2.28, y = -5.55$
16 $x = 2.83, y = -2.83$ or $x = -2.83, y = 2.83$
17 $x = 1.28, y = 4.83$ or $x = -1.88, y = -4.63$
18 $x = 0.16, y = 3.16$ or $x = -3.16, y = -0.16$

Revision exercise C1 (page 297)

1 a) $\dfrac{5x + 4}{6}$ b) $\dfrac{2x - 17}{20}$

c) $\dfrac{3x}{(x + 1)(x - 2)}$ d) $\dfrac{x^2 + 6x - 1}{(x - 1)(x + 2)}$

e) $\dfrac{3x}{x + 1}$

2 a) $x = 3$ b) $x = 1\frac{5}{7}$
c) $x = -5$ or 3 d) $x = 4$ or -3
e) $x = -2$ f) $x = 3$ or -5
g) $x = 4$ or 5

3 a) $\begin{pmatrix} 2 \\ 4 \end{pmatrix}$ b) $\begin{pmatrix} 3 \\ 1 \end{pmatrix}$ c) $\begin{pmatrix} 2 \\ -2 \end{pmatrix}$

d) $\begin{pmatrix} -3 \\ 4 \end{pmatrix}$ e) $\begin{pmatrix} 1 \\ 0 \end{pmatrix}$ f) $\begin{pmatrix} 0.5 \\ 1 \end{pmatrix}$

g) $\begin{pmatrix} 5 \\ 13 \end{pmatrix}$ h) $\begin{pmatrix} -0.5 \\ 2 \end{pmatrix}$ i) $\begin{pmatrix} 3.5 \\ 2.5 \end{pmatrix}$

4 a) $(-1, 3)$ b) $(0, 0)$
c) $(5, -5)$

5 $\overrightarrow{BC} = 2\mathbf{b} - \mathbf{a}$

6 a) $\overrightarrow{AB} = 2\mathbf{b} - 2\mathbf{a}$, $\overrightarrow{BC} = 4\mathbf{b} - 4\mathbf{a}$
b) ABC is a straight line and BC is twice the length of AB.

7 a) $\overrightarrow{EB} = \frac{1}{2}\mathbf{p}$, $\overrightarrow{BF} = \frac{1}{2}\mathbf{q}$, $\overrightarrow{EF} = \frac{1}{2}\mathbf{p} + \frac{1}{2}\mathbf{q}$, $\overrightarrow{HD} = \frac{1}{2}\mathbf{q}$, $\overrightarrow{DG} = \frac{1}{2}\mathbf{p}$, $\overrightarrow{HG} = \frac{1}{2}\mathbf{p} + \frac{1}{2}\mathbf{q}$

b) HG and EF are equal and parallel.

8 $\overrightarrow{EB} = \frac{1}{3}\mathbf{p} - \frac{2}{3}\mathbf{q}$

9 $\overrightarrow{DC} = 5\mathbf{p}$. So AB and DC are parallel. So the shape is a trapezium.

10 Median = £212 000, IQR = £30 000

House prices are higher in the south-east, with a wider spread of prices.

11 a)

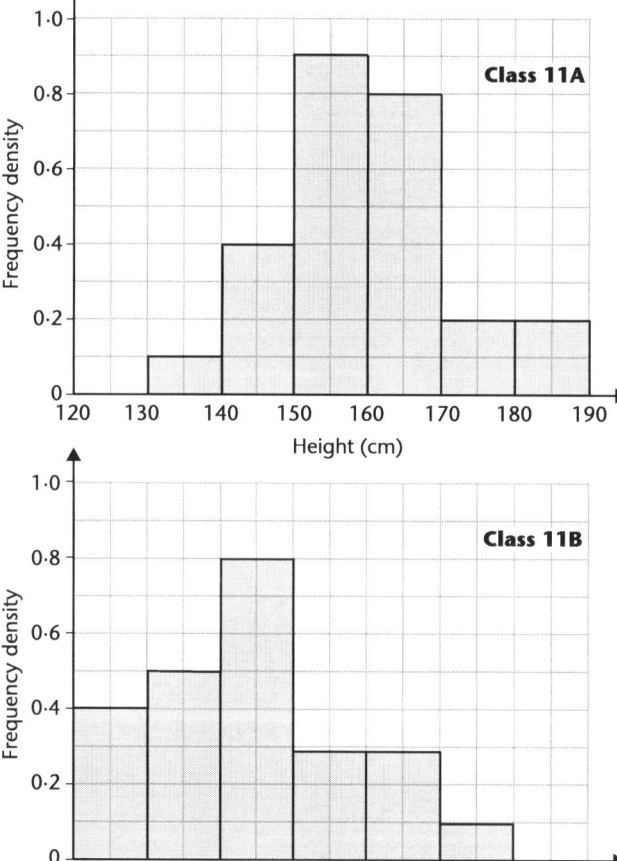

b) French: median 44, IQR 16 English: median 48, IQR 29 The English marks are higher but with a wider spread.

12 a)

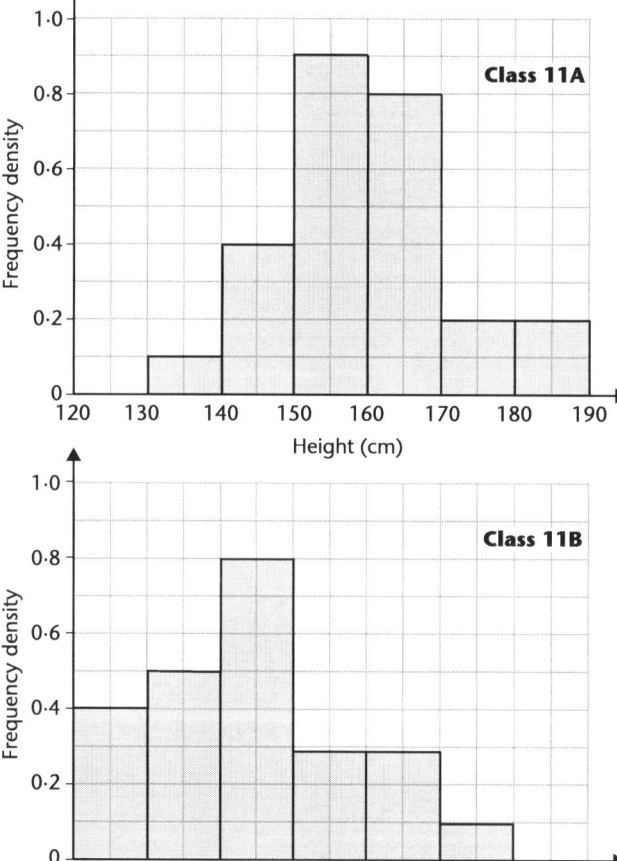

b) The students in class 11A are taller on average than those in class 11B.
The range of the heights is approximately the same for both classes but class 11A has a smaller interquartile range.

STAGE
10

13 a) $x = 1, y = 2$ or $x = 3, y = 6$
b) $x = 1, y = 2$ or $x = 2, y = 5$
c) $x = 1, y = 2$ or $x = -1, y = 10$
d) $x = 0, y = 6$ or $x = -6, y = 0$
14 a) $x = -2, y = 10$ or $x = 4, y = 4$
b) $x = \frac{1}{2}, y = -\frac{1}{2}$ or $x = 3, y = 7$
c) $x = 3, y = -1$ or $x = 2\frac{1}{2}, y = -1\frac{1}{4}$
d) $x = 5, y = -2$ or $x = -4\cdot6, y = 2\cdot8$
15 $x = -2\cdot9, y = -0\cdot9$ or $x = 0\cdot9, y = 2\cdot9$

13 Trigonometrical functions

Exercise 13.1 (page 304)

1 Check students' graphs.
A sketch is given here as a guide.

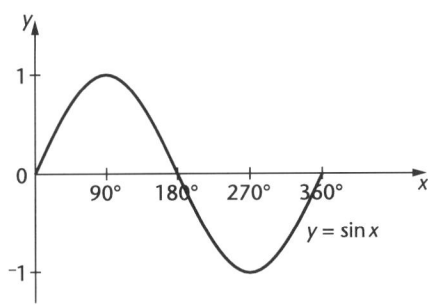

a) $x = 0°, 180°, 360°$
b) $x = 27°, 153°$
c) $x = 217°, 323°$

2 Check students' graphs.
A sketch is given here as a guide.

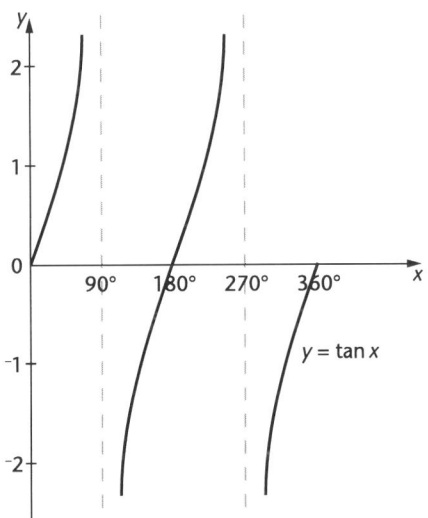

a) $x = 45°, 225°$
b) $x = 63°, 243°$

3

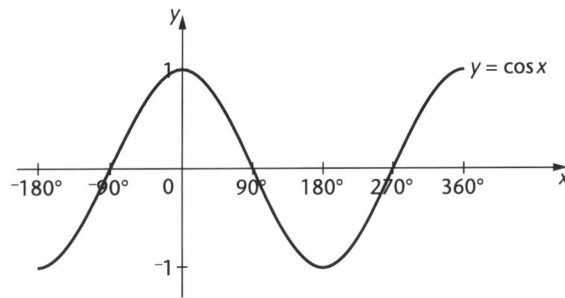

$x = \pm72\cdot5°, 287\cdot5°$
4 See sketch for question **2**.
$x = 117°, 297°$
5

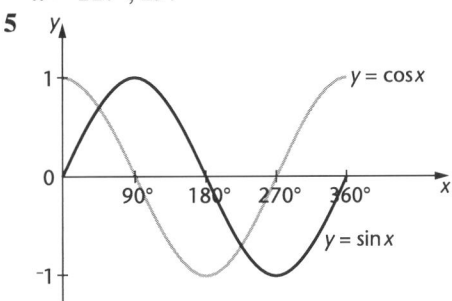

$x = 45°, 225°$
6 $x = 120°, 240°, 480°, 600°$
7

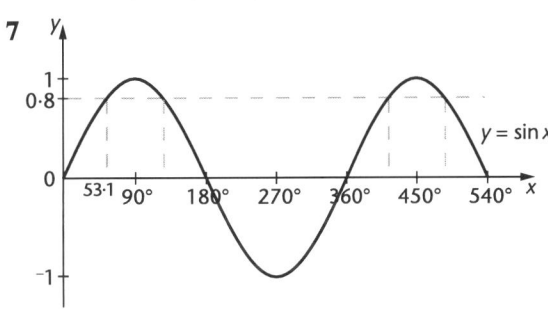

$x = 53\cdot1°, 126\cdot9°, 413\cdot1°, 486\cdot9°$
8 See sketch for question **1**.
$x = 192°, 348°$
9 $x = 253°, 467°$
10 $x = 135°, 315°$

11 a)

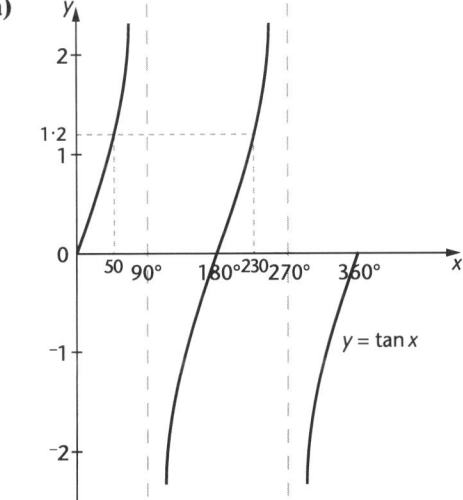

b) $x = 50°, 230°$

12 e.g. **a)** 160°, 380°, 520°   **b)** 60°, 420°, 480°
      c) 225°, 315°, 585°   **d)** 30°, 150°, 510°
      e) 140°, 400°, ⁻220°   **f)** ⁻100°, 260°, 280°
      g) 50°, 410°, 490°   **h)** ⁻60°, 240°, 300°

13 e.g. **a)** 220°, 500°, 580°   **b)** 120°, 240°, 480°
      c) ⁻40°, 320°, 400°   **d)** 270°, 450°, 630°
      e) 75°, 435°, 645°

14 e.g. **a)** ⁻135°, 225°, 405°   **b)** ⁻60°, 300°, 480°
      c) 220°, 400°, ⁻140°   **d)** 100°, 280°, 460°
      e) 210°, 390°, 570°   **f)** ⁻45°, 315°, 495°

15 **a)** $x = 201{\cdot}7°, 338{\cdot}3°$ **b)** $x = 0°, 360°$
      c) $x = 0°, 180°, 360°$ **d)** $x = 30°, 150°$
      e) $x = 122{\cdot}7°, 302{\cdot}7°$ **f)** $x = 210°, 330°$
      g) $x = 45°, 225°$ **h)** $x = 48{\cdot}2°, 311{\cdot}8°$

Exercise 13.2 (page 307)

In questions **1** to **4**, check students' graphs.
Sketches are given here as guides.

1

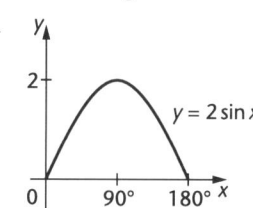

2

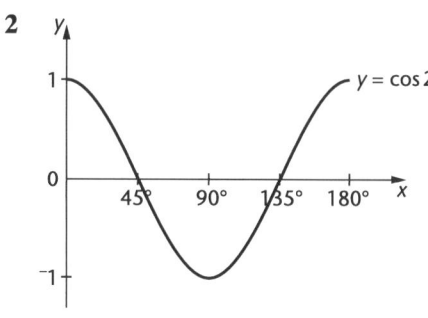

3

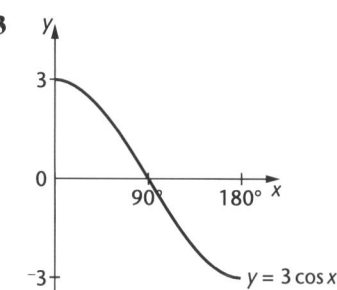

4

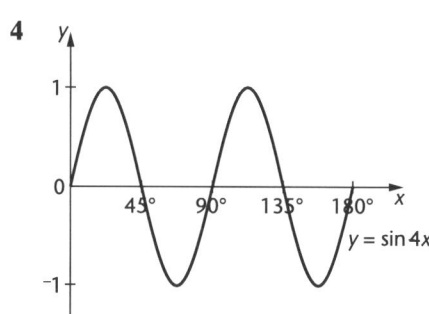

5 **a)** 3, 360°   **b)** 4, 180°
    c) 2, 720°   **d)** 5, 360°
    e) 2, 120°   **f)** 4, 1080°

6 $a = 5$ $b = 0{\cdot}5$

7 $a = 0{\cdot}5$ $b = 3$

8 $a = 2$ $b = 3$

9 $a = 3$ $b = 0{\cdot}5$

10

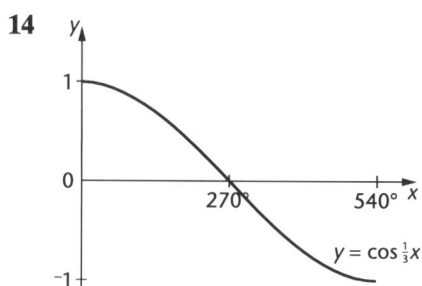

11 $x = 15°, 75°, 195°, 255°$

12 $x = 45°, 135°, 225°, 315°$

13

One solution

14

15 $x = 60°, 180°, 300°$

16 $x = 30°, 150°$

17

Two solutions

14 Transforming functions

Exercise 14.1 (page 313)

1 a)

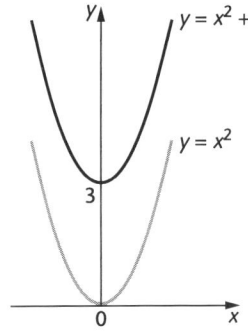

$y = x^2 + 3$

$y = x^2$

b) Translation of $\begin{pmatrix} 0 \\ 3 \end{pmatrix}$

2 a)

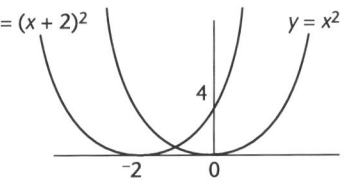

$y = (x + 2)^2$ $y = x^2$

b) Translation of $\begin{pmatrix} -2 \\ 0 \end{pmatrix}$

3 a)

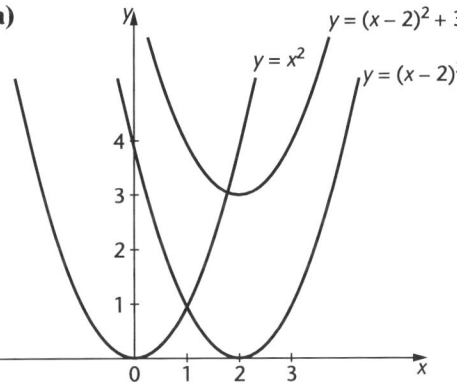

$y = (x - 2)^2 + 3$
$y = x^2$
$y = (x - 2)^2$

b) Translation of $\begin{pmatrix} 2 \\ 3 \end{pmatrix}$

4 a)

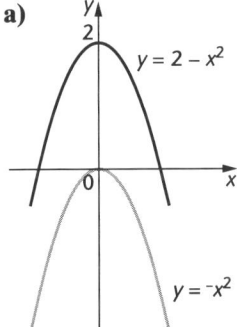

$y = 2 - x^2$

$y = -x^2$

b) Translation of $\begin{pmatrix} 0 \\ 2 \end{pmatrix}$

5 a)

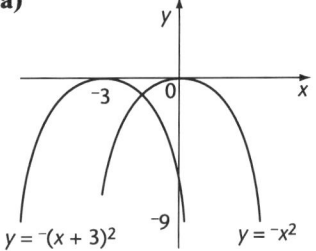

$y = -(x + 3)^2$ $y = -x^2$

b) Translation of $\begin{pmatrix} -3 \\ 0 \end{pmatrix}$

6 a)

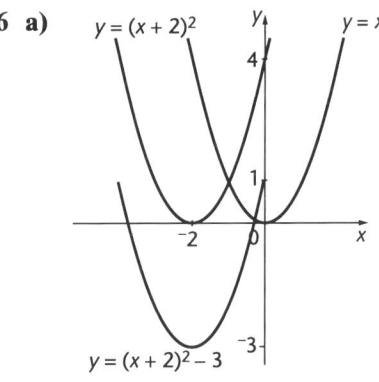

$y = (x + 2)^2$ $y = x^2$

$y = (x + 2)^2 - 3$

b) Translation of $\begin{pmatrix} -2 \\ -3 \end{pmatrix}$

7 a)

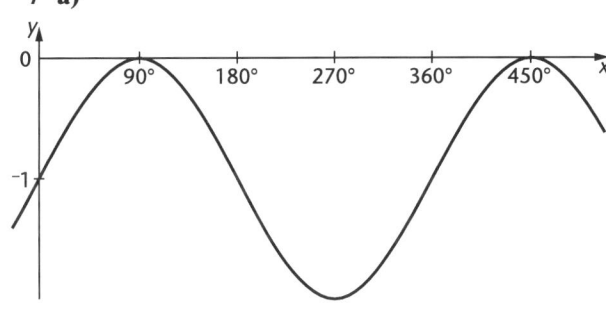

b) $y = \sin x - 1$

8 a)

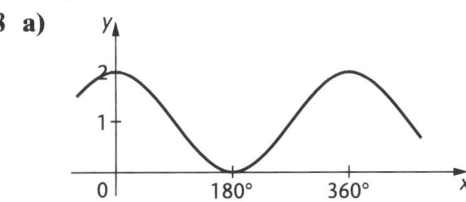

b) $y = \cos x + 1$

9 a) $y = \sin(x - 3)$ **b)** $y = \sin x + 4$

10 a) $y = x^2 - 5$ **b)** $y = (x + 2)^2$
c) $y = (x - 1)^2 + 2$ **d)** $y = (x - 3)^2 - 4$

11 a)

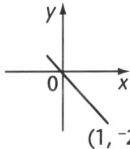

$(1, -2)$

b)

y
6

0   3   x

12 a) $(1, 3)$ **b)** $(2, 4)$

13 a) $(1, 0)$ **b)** $(^-1, 2)$

14 $y = \cos x + 2$

15 a) $y = (x + 2)^2 + 3$

   b) Check students' answers.

16 $y = \sin x - 1$

17 a)

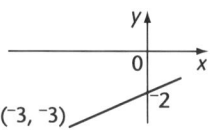

   b)

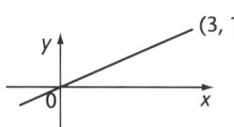

18 a) Check students' answers.

   b) $(3, ^-8)$

Exercise 14.2 (page 318)

1 a)

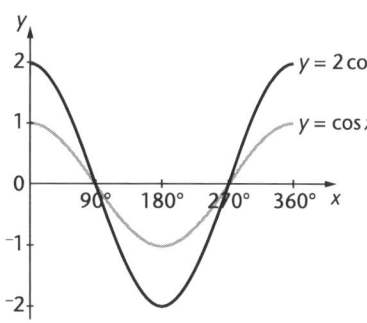

   b) One-way stretch parallel to y-axis with scale factor 2.

2 a)

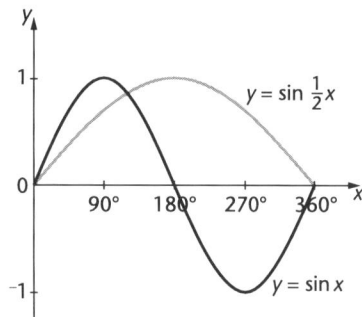

   b) One-way stretch parallel to x-axis with scale factor 2.

3 One-way stretch parallel to x-axis with scale factor $\frac{1}{3}$.

4 a) Reflection in the y-axis.

   b) Reflection in the x-axis.

   c) One-way stretch parallel to the y-axis with scale factor 3.

5 a)

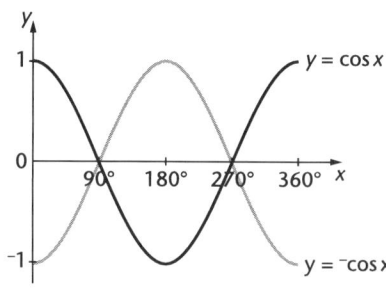

   b) Reflection in the x-axis.

6 a)

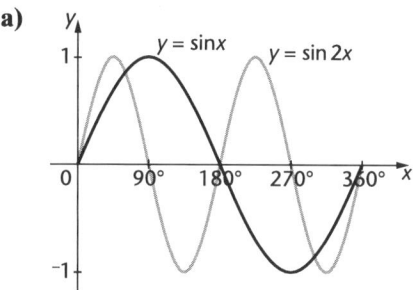

   b) One-way stretch parallel to x-axis with scale factor $\frac{1}{2}$.

7 One-way stretch parallel to x-axis with scale factor 3.

8 a) Reflection in the x-axis.

   b) Reflection in the y-axis.

   c) One-way stretch parallel to the y-axis with scale factor 5.

9 $y = \cos 3x$

10 a) $y = x^2 + 5$

   b) $y = ^-x^2 - 5$

11 a) $y = 3x + 6$

   b) $y = 2x + 2$

12 a) Translation of $\begin{pmatrix} 0 \\ 1 \end{pmatrix}$.

   b) One-way stretch parallel to the y-axis with scale factor 3.

   c) One-way stretch parallel to the x-axis with scale factor $\frac{1}{2}$.

   d) One-way stretch parallel to the x-axis with scale factor $\frac{1}{3}$ and one-way stretch parallel to the y-axis with scale factor 5.

13 a) $y = ^-2x - 1$

   b) $y = ^-2x + 1$

   c) $y = 4x + 1$

14 a) $y = \frac{1}{4}x^2$

   b) $(2, 1)$

   c) (i) $\frac{1}{4}$ **(ii)** $\left(1, \frac{1}{4}\right)$

15 $y = \sin 4x$

16 a) $y = x^2 - 1$

   b) $y = ^-x^2 + 1$

17 a) $y = 16x + 4$

   b) $y = 8x + 1$

Graduated Assessment for OCR GCSE Mathematics © Hodder Murray 2007

STAGE
10

18 a) Translation of $\begin{pmatrix} 0 \\ -2 \end{pmatrix}$.

b) One-way stretch parallel to the y-axis with scale factor 3.

c) One-way stretch parallel to the x-axis with scale factor 2.

d) One-way stretch parallel to the x-axis with scale factor $\frac{1}{2}$ and one-way stretch parallel to the y-axis with scale factor 4.

19 a) $y = {}^-x^2 - 3$
b) $y = x^2 + 3$
c) $y = 4x^2 + 3$

20 a) $y = x^2 - 2x$
b) $y = {}^-x^2 - 2x$
c) $y = {}^-(x - 3)^2 + 2(x - 3)$ or $y = {}^-x^2 + 8x - 15$

21

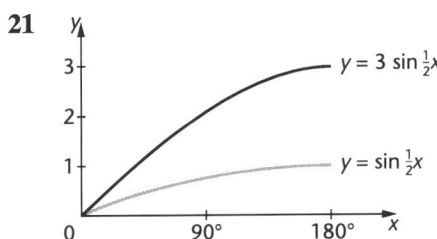

22 $a = 3$
23 $b = \frac{1}{4}$
24 $a = 3, b = 6$
25 $a = {}^-1$

15 Probability

Exercise 15.1 (page 324)

1 a) 0·195 to 3 d.p.
b) 0·519 to 3 d.p.

2 a) $\frac{25}{49}$ **b)** $\frac{10}{49}$ **c)** $\frac{45}{49}$

3 a) 0·382 to 3 d.p.   **b)** 0·618 to 3 d.p.
c) 0·891 to 3 d.p.

4 a) 0·275 to 3 d.p.   **b)** 0·123 to 3 d.p.
c) 0·444 to 3 d.p.   **d)** 0·718 to 3 d.p.

5 a) $\frac{3}{7}$ **b)** $\frac{4}{7}$ **c)** $\frac{3}{7}$

6 0·027

7 a) $\frac{3}{100}$ **b)** $\frac{3}{99} = \frac{1}{33}$
c) $\frac{2}{98} = \frac{1}{49}$ **d)** $\frac{6}{970\,200} = \frac{1}{161\,700}$

8 a) 0·059 to 3 d.p.   **b)** 0·013 to 3 d.p.
c) 0·414 to 3 d.p.   **d)** 0·586 to 3 d.p.

9 a) $\frac{7}{12}$ **b)** $\frac{6}{11}$
c) 0·1515

10 a) $\frac{4}{15}$ **b)** $\frac{7}{15}$

11 a) 0·497 to 3 d.p.   **b)** 0·477 to 3 d.p.
c) 0·222 to 3 d.p.

12 a) 0·24   **b)** 0·86

13 a) $\frac{2}{35}$ **b)** $\frac{13}{35}$

14 0·225

Revision exercise D1 (page 327)

1

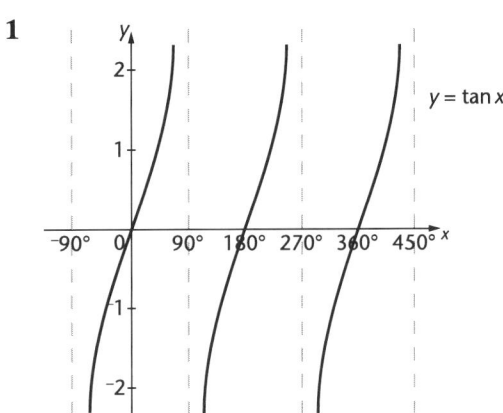

2 $x = 45°, 225°$

3

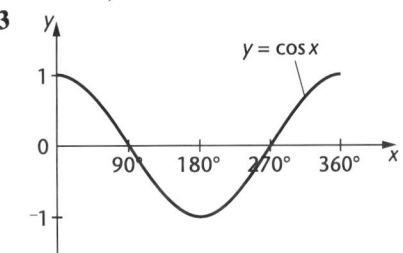

$x = 217°$

4 $x = 210°, 330°$

5 $x = 78·5°, 281·5°$

6

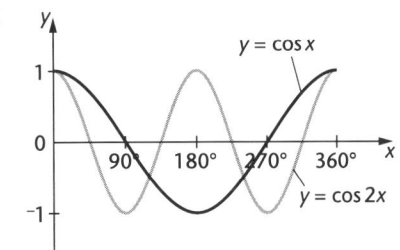

7 $x = 45°, 225°$

8 A(60, 0), B(30, 2)

9

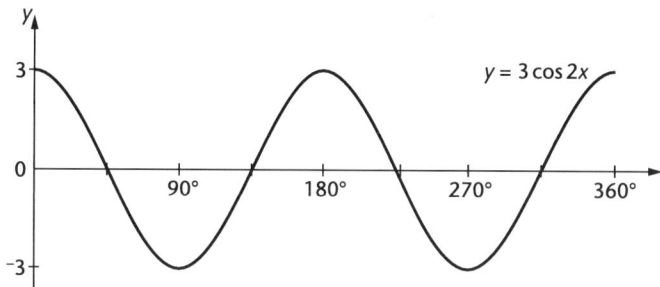

10 $y = (x - 1)^2 - 2$ or $y = x^2 - 2x - 1$

11 a) One-way stretch parallel to the x-axis with scale factor $\frac{1}{3}$.
  b) One-way stretch parallel to the y-axis with scale factor 4.
  c) Reflection in the y-axis.

12

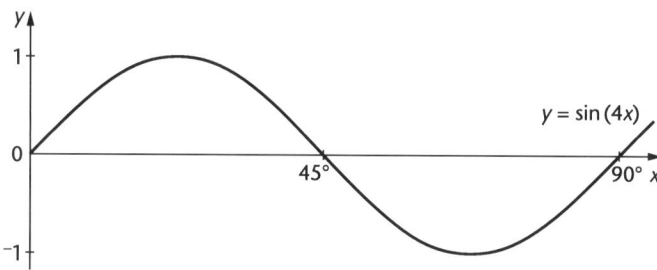

13 a)

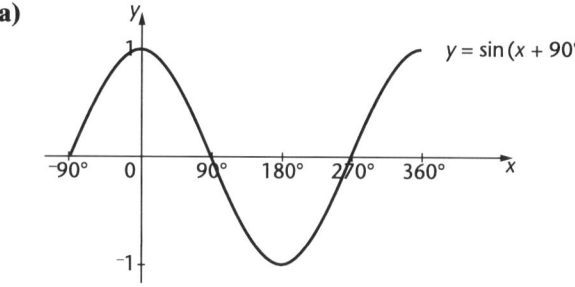

  b) $y = \cos x$

14 a) $y = \cos x + 3$
  b) $y = \cos 4x$

15 a) $y = 2[(x - 3)^2 + 1]$ or equivalent e.g. $y = 2x^2 - 12x + 20$
  b) $(3, 2)$

16 a) $\frac{1}{19}$ **b)** $\frac{5}{76}$ **c)** $\frac{21}{38}$

17 a) $0 \cdot 056$ **b)** $0 \cdot 332$ **c)** $0 \cdot 612$

STAGE
10

Mathematical grids

STAGES
9/10

Formulae sheet

For the exact content of the formulae sheets to be provided in the examinations, please check with the examination board.

Higher tier

Volume of a prism = area of cross-section × length

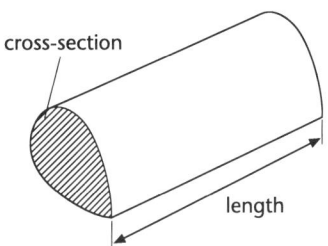

For any triangle ABC

Sine rule: $\dfrac{a}{\sin A} = \dfrac{b}{\sin B} = \dfrac{c}{\sin C}$

Cosine rule: $a^2 = b^2 + c^2 - 2bc \cos A$

Area of triangle $= \frac{1}{2}ab \sin C$

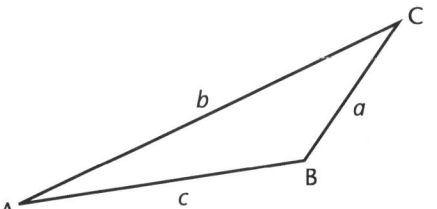

Volume of a sphere $= \frac{4}{3}\pi r^3$

Surface area of a sphere $= 4\pi r^2$

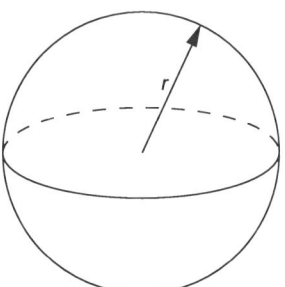

Volume of a cone $= \frac{1}{3}\pi r^2 h$

Curved surface area of a cone $= \pi r \ell$

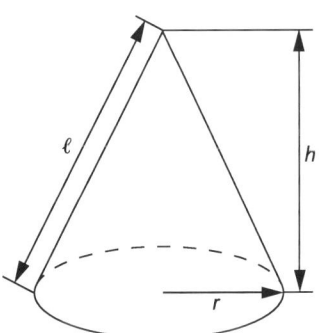

The quadratic formula

The solution of the equation $ax^2 + bx + c = 0$, where $a \neq 0$, is given by

$$x = \frac{-b \pm \sqrt{b^2 - 4ac}}{2a}$$

STAGES
9/10